TRANSACTIONS

OF THE

AMERICAN PHILOSOPHICAL SOCIETY

HELD AT PHILADELPHIA
FOR PROMOTING USEFUL KNOWLEDGE

NEW SERIES—VOLUME 64, PART 4
1974

MAPPAE CLAVICULA
A LITTLE KEY TO THE WORLD OF MEDIEVAL TECHNIQUES

CYRIL STANLEY SMITH
Massachusetts Institute of Technology
and
JOHN G. HAWTHORNE
University of Chicago

THE AMERICAN PHILOSOPHICAL SOCIETY
INDEPENDENCE SQUARE
PHILADELPHIA

July, 1974

Library of Congress Catalog
Card Number 74-77910
International Standard Book Number 0-87169-644-4
US ISSN 0065-9746

MAPPAE CLAVICULA

A LITTLE KEY TO THE WORLD OF MEDIEVAL TECHNIQUES

An annotated translation based on a collation of the Sélestat and Phillipps-Corning manuscripts, with reproductions of the two manuscripts

CYRIL STANLEY SMITH and JOHN G. HAWTHORNE

CONTENTS

INTRODUCTION

THE MANUSCRIPTS AND THEIR BACKGROUND

There are few conventional written records that have survived from the early Middle Ages in Europe. Both scholarship and commerce were at a low ebb, but there was plenty of creative activity in the field of art, and also, as has become increasingly apparent in recent years,[1] in technology. The practical people who were developing stirrups, substituting wind and water power for man's labors, and enlarging the scale of metallurgy wrote virtually nothing, and even if they had done so their notes would not have seemed important to the custodians of state or ecclesiastical documents. The evidence for the technology lies mainly on the walls and in the treasure rooms of ecclesiastical buildings and in the objects uncovered by archaeologists' spades. There are, however, a few literary remains and these, of course, deserve the closest scrutiny. Most of the people who have discovered and studied the manuscripts have done so for linguistic or philological reasons, but they were made as records of technology and the relationship between the words preserved in the libraries and the operations actually being carried out in workshops provides a fascinating problem.

Not a single original note on the technology of materials has survived in any European language from the period of half a millennium following the Alexandrian Greek papyri now in Leyden and Stockholm and presumably dating from the third century A.D.[2] These consist of simple compilations of recipes for making colored metals and artificial gems, for dyeing fabrics, and for a few other related operations.

The literary tradition did, however, somehow continue and it comes to the surface again around the year 800 with the appearance of two manuscripts in much the same form as the Alexandrian ones but now in Latin, not Greek, and with a somewhat wider range of topics included. The first one to be studied by a modern scholar is a late-eighth- or early-ninth-century manuscript at Lucca where it is associated with the *Liber Pontificalis* in Codex Lucensis 490. This was published in 1739 by the eminent Italian scholar Ludovico Antonio Muratori in his essay on Italian art after the fall of Rome which appears in the second volume of his sumptuous *Antiquitates Italicae medii aevi*. It has since been repeatedly analyzed by historians of chemistry for its technical content as well as by paleographers, philologists, and latinists.[3]

---

[1] See especially Lynn White, Jr., *Medieval Technology and Social Change* (Oxford, 1962).

[2] Marcellin Berthelot has more than anyone else shown the importance of these for the history of chemistry. For the text see Berthelot (1885, 1899) and Lagercrantz (1913), and for an English translation based on Berthelot's French see Caley (1926, 1927). [*Note:* A name followed by a date is to be regarded throughout this work as a reference to the complete bibliography on pages 120 to 122.]

[3] See the bibliography for editions by Muratori (1739), Duchesne (1886), Burnam (1920), Schiaparelli (1924), and Hedfors (1932). We have been unable to obtain a copy of Schiaparelli, which is a complete photocopy of the manuscript. Hedfors includes a German translation and extensive notes: it is by far the best edition. Burnam's edition, with English translation, is inferior. Svennung (1941) analyzed the grammatical style of the Lucca manuscript in detail. We have found his discussion of the origin, mutation, and transmission of many words to be of great help in dealing with difficult phrases in the *Mappa*.

The Lucca manuscript, called by Muratori *Compositiones ad tingenda musiva* but more properly called *Compositiones variae* (R. P. Johnson [1934–1935]), is, however, only one member of what is actually a series of medieval Latin manuscripts containing recipes for making pigments, dyes, and colored metals, and for other operations relating to the practical crafts. Contemporary with the Lucca manuscript, but north of the Alps instead of in Italy, there was transcribed another compilation which includes virtually everything that is in Lucca plus many additional recipes. This is the *Mappae Clavicula*. Of this there exist a fragment from the early ninth century, an extended manuscript of the tenth century, and the most complete one dating from the twelfth century, which we translate here. In addition to these there are the quite separate works of Eraclius, *De coloribus et artibus Romanorum* (the earliest of which is also tenth century) and in the early twelfth century the superb treatise *De diversis artibus* by the pseudonymous Theophilus. The last is of a quite different character from the other manuscripts, being not a compilation but an original treatise written by a man in the workshop himself writing from first-hand knowledge. By the twelfth century, however, scribes were busily copying anything that seemed of interest, and there are innumerable manuscripts which contain portions of all of these in various different arrangements and with decreasing accuracy.

There was a remarkable flurry of interest in the literary history of painting in the 1830's, especially in England under the leadership of Charles Eastlake. Most productive was the remarkable Mary Philadelphia Merrifield who was commissioned by the British government to travel to Italy for the purpose of collecting and studying manuscripts "with a view principally of ascertaining the processes and methods of oil painting adopted by the Italians." The decade also saw the publication (among several later works) of two editions of Theophilus, one of Eraclius and the first appearance of the twelfth-century *Mappae Clavicula*, which was published by its owner, the famed book collector Sir Thomas Phillipps, in 1847. Despite the fact that the *Mappae Clavicula* contains a good deal more material of technical interest than does the *Compositiones variae*, no other edition of it has been published. It is well known for its pigment recipes but it has been inadequately explored by historians of other aspects of technology, and it is hoped that the present translation will serve to increase interest both in the manuscript itself and in the state of technology that it reflects.

The earliest dated evidence for the existence of a work entitled *Mappae Clavicula* appears to be the entry reading "Mappae Clavicula de efficiendo auro volumen 1" in a catalog of the library of the Benedictine monastery of Reichenau on Lake Constance that was made in the years 821–822 (Johnson, 1935;

Holder, 1916; Lehmann, 1918). The manuscript itself is no longer in existence. As Berthelot (1893) suggested, however, it is likely that even this manuscript was itself a copy of one or more earlier ones. This view is based solely on indirect and comparative evidence, but the traditional nature of the recipes, many of which go back for centuries, the several different styles of both the processes described and their language, and the fact that this was an age of widespread copying and international exchange of knowledge makes it highly probable.

In 1941 Wilhelm Ganzenmüller described a two-leaf parchment fragment of the *Mappae Clavicula* which had been found in the library of the Augustine Choristers Foundation at Klosterneuburg (near Vienna) by Herr Černik and dated by him, on paleographic evidence, as first third of the ninth century. Mr. B. J. Hennessey of the Department of Classics, Harvard University, to whom the present translators showed the facsimile published by Ganzenmüller independently commented that the writing "seems to be very early Carolingian, most probably from the late eight or early ninth century. Not enough survives to form a very good idea of the script. The letter forms on these leaves are typical and widespread. No ligatures are used and only one abbreviation (scđm) which again is known throughout Europe. It could have been produced in the Northern parts of France, Germany, or Belgium."

The Klosterneuburg fragment is therefore virtually contemporary with the Lucca manuscript, but the fact that only four of the seventeen recipes on the two sides of folio 1 are in Lucca, while all of them are in the tenth and twelfth-century versions of the *Mappae Clavicula* places it squarely at the head of the latter tradition. Folio 2 of the Klosterneuburg fragment contains material common to all. The fact that Lucca is included in the *Mappa* and not *vice versa* suggests, of course, that the *Mappa* is the later compilation, though Lucca could perhaps be an excerpt from a larger work. All one can say with confidence is that both traditions were already in existence in Europe at the very beginning of the ninth century—the *Compositiones variae* south of the Alps, the *Mappae Clavicula* north of them.

THE SÉLESTAT MANUSCRIPT

With the Sélestat manuscript—which is reproduced in its entirety in Appendix A—we are at last on firm ground. It is the earlier of two existing comprehensive manuscripts of the *Mappae Clavicula* and forms part of a volume preserved in the Library of the town of Sélestat (Bas-Rhin). Designated MS 17 (previously Latin 360), it consists of 215 parchment leaves, 14.6 × 19.1 cm., and includes also Vitruvius *de Architectura* (which constitutes the longest part of the manuscript) and miscellaneous other material on building and other topics. It was discovered and described by A. Giry (1878) who dated it as tenth century, a date accepted both by Berthelot (1893) who used it extensively and Johnson (1937) who briefly examined it. Mr. Barry Hennessey has studied a photocopy for the present translators and writes as follows:

The Sélestat ms. is written in a carolingian minuscule of the early ninth century. One can be less certain about its place of origin. It was likely St. Amand, near Tours, in northern France. E. A. Lowe (introduction to *Codices Latini Antiquiores*, vol. 10) lists eleven features which characterize the St. Amand script, advising that several of them, but never all, appear together in any one ms. Four of these features are found in the Sélestat: the shafts of the *s* and *r* descending well below the line; the tall *T* with sinuous top in ligature with *N*; the abbreviation of *est* and *esse* as *ē* and *ēē*; and the long tail on the *R* in the *OR* ligature, sweeping below the line. However, many mss. of the St. Amand style were written at Salzburg, Austria. Arno, abbot of St. Amand 783–821, was also bishop of Salzburg 785–821, and many scribes from St. Amand travelled to Salzburg and worked in the scriptorium there. If the ms. had been in Austria or Germany at an earlier time, Salzburg would be its likely origin, but its later provenance strongly suggests the St. Amand scriptorium.

We have worked with photocopies graciously provided by the librarian of the Sélestat town library and the Centre National de la Recherche Scientifique. As can be seen from our Table of Concordance, pp. 10–14, the Sélestat manuscript includes virtually all of the Lucca recipes and adds important new ones, but it is only approximately half as long as the twelfth-century version now to be described.

THE PHILLIPPS-CORNING MANUSCRIPT

This manuscript, widely known as Phillipps MS 3715, is now in the Corning Museum of Glass at Corning, New York. A complete reproduction forms Appendix B of the present edition. This is the most comprehensive of the manuscripts of *Mappae Clavicula* and is the best known, for reasons that are not far to seek: it is the only one that has been published. This was done by its owner, Sir Thomas Phillipps in 1847.

The Phillipps manuscript is in the form of a small volume, 17.0 × 11.9 cm., written in a clear hand on vellum, bound in old calf (not of course contemporary). It contains 67 leaves, most with 21 lines to the page, with the spacing marked by marginal prickings or pounce marks. The leaves are numbered in Arabic numerals, obviously not by the original scribe. There are no colored decorations, but the chapter initials, and the rubrics for the chapter titles where they are inserted alternate in red and green. (See frontispiece, which is a natural-size reproduction of folio 18). Although occasionally the rubricator erred, this chromatic alternation is the best indication of the intended division into chapters. The early chapters carry Roman enumeration in the original hand, but

after No. LXVI the chapters were without numbers until some were assigned rather arbitrarily by Phillipps. Many of the chapter titles were also omitted, although spaces had usually been left for them, evidently with the intention of later rubrication.[4] When the spaces that had been left were inadequate, the titles often extended into the margin or into spaces between the regular lines.

The life of Sir Thomas Phillipps (1792–1872) and his amazing career as an almost maniacal collector of manuscripts has been described in interesting detail by A. N. L. Munby in his *Phillipps Studies* (5 v., Cambridge, 1951 to 1960), summarized in his *Portrait of an Obsession* (London, 1967). Phillipps had a passion for historical manuscripts, though his huge collection of over 60,000 items was made with little principle or perspective. Something as obviously interesting as the *Mappae Clavicula* would doubtless have been preserved in some other library had Phillipps not bought it, but the very existence of the good market created by Phillipps's activities undoubtedly saved many other manuscripts from the wastepaper dealer. After Phillipps's death on 6 February, 1872, the collection passed to his son-in-law, Thomas Fitzroy Fenwick, who sold many items at a series of important auction sales that created great excitement in museum and library circles. After his death in 1938, the unsold items became the property of his nephew Alan George Fenwick, but World War II prevented further disposal. Early in 1946 the entire residue of Phillipps's collection, which was still enormous, passed in a courageous purchase for the sum of £100,000 into the possession of the booksellers William H. Robinson Ltd. of Pall Mall. From them, in October, 1952, Mr. Arthur Houghton purchased the manuscript of the *Mappa* by private treaty on behalf of the Corning Museum of Glass, where it now is. It became water-soaked when the museum was inundated during the disastrous flood that occurred in upper New York State on 23 June, 1972, but it was retrieved and immediately transferred to a deep-freezing unit to preserve it. As we go to press (January, 1974) the manuscript still awaits vacuum drying, but Mrs. Carolyn Horton reports that it is in good condition, with the vellum unharmed and the red and dark-brown text perfectly legible. She predicts that the only damage from the flood that will be visible after restoration is complete will be some

---

[4] Many of the rubrics in folios 6 to 13 in Sélestat are similarly lacking, together with the initials. Such omissions are of course common in medieval manuscripts, for the rubrication was done later than the main text and often by a different scribe. In his published transcript Phillipps frequently failed to follow the rubricated initials and titles, and, reading titles as text, he has collapsed two or more of the original chapters into one. This suggested to us the possibility that Phillipps may have been color blind: indeed when using a photocopy of the manuscript we ourselves made the same errors and only arrived at the correct division of chapters on the basis of the unmistakable black, green, and red of the original.

The ownership stamp of Jacques Antoine Rabaut-Pomier (1744–1820). From folio 17r of the Phillipps-Corning manuscript.

transfer of the blue initials (written in folium?) on to the opposing vellum surface. The photographs which form the frontispiece and appendix B of the present edition were prepared before the flood.

Sir Thomas Phillipps had acquired the manuscript of the *Mappae Clavicula* in 1824 from the Rev. Mons. Allard in Paris. Joseph-Félix Allard was born in Provence at an unknown date and after taking orders became professor of rhetoric at a small seminary in Marseilles. He published a translation of Tertullian's *Apologeticum* in 1827. To enable him to pursue his literary studies in greater depth he was released by the bishop of his own district and became attached to the parish of St. Eustache in Paris, but he died on the twenty-eighth of October, 1831, when he was hardly more than thirty years old. (*Dictionnaire de Biographie Française* [Paris, 1936] 2: p. 115.)

Prior to this the manuscript had belonged to Jacques Antoine Rabaut-Pomier, whose stamp appears on several of the leaves. Rabaut-Pomier (1744–1820) was a pastor of the French protestant church who held minor public offices during and after the Revolution. He claimed, apparently with justice, to have suggested vaccination in 1784 as a result of an earlier observation of the immunity of men who had contracted mild disease from handling pustulated cattle. No historical publications are attributed to him and neither he nor any earlier owner seem to have done anything with the information in the *Mappa*.

Phillipps published the entire text of his manuscript in the journal *Archaeologia* in 1847, and this version is all that has hitherto been available to scholars. The background of this publication and the important part played by Albert Way, Director of the Society of Antiquaries at the time, is discussed below (pp. 7 to 9).

On the origin of his manuscript, Phillipps observed that the hand appears to be that used in Flanders or England at the time of Henry II (r. 1154–1189) but adds his view that "the presumption will be . . . in favour of England." This he bases on the use of two English words "gatetriu" (goat tree) and "greningpert" (greningwert = greningwort) in chapters 190 and 191. To this he could have added the fact that the tables of runes (Chap. 288B and 288P) follow the Anglian system. However, this does not imply that the author of the *Mappae Clavicula* was an Englishman, only that one in the long series of

transcribers may have been. Phillipps could equally well have argued that the author was from the Middle East, for many Arabic terms appear in chapters 195 to 201. There was, of course, no single author but a whole sequence of contributors, editors, and copyists. Note the frequent use throughout of latinized forms of Greek words which indicate the very ancient classical origin of some of the recipes, the Byzantine origin of others, and the intrusion of traditional recipes from many sources into the rather polyglot Latin vocabulary of the time.

R. W. Hunt and N. Kerr of Oxford University studied a photocopy of some pages of the manuscript for Roosen-Runge and were sure that the writing was not English, while Dr. M. G. I. Lieftinct of Leyden thought it was possibly northwest France, *ca.* 1170 (Roosen-Runge, 1967: 1: p. 70). Quite independently, Mr. B. J. Hennessey of Harvard University, Department of Classics, examined our photocopies and reports as follows:

The script of Phillipps 3715 is a late carolingian minuscule, already showing many features of the gothic style, which appeared in the thirteenth century. It is unfortunately very difficult to date manuscripts from this period, since no precise criteria have been established. Unpublished studies in the files of Dr. Peter Elder dealing with dated French manuscripts of the eleventh and twelfth centuries yield a few general guidelines which suggest a mid-twelfth-century date for the Phillipps manuscript. The fairly round bows in the b and d, the prominent hairlines at the foot of the vertical shafts, the use of hyphens, the clubbing of the b, h, and l, and the regular use of uncial a—all these are most commonly found in manuscripts of the early twelfth century, although they certainly also occur in the eleventh and late twelfth centuries. I know of no attempt that has yet been made to establish a pattern in the abbreviations and ligatures used.

With regard to the place of origin, again I am vague. The same script was employed all over northern Europe, with little of what might be called distinct local styles. The one manuscript I found most closely resembling the Phillipps comes from Rouen, but is dated 1221, quite later than I would have expected. Germany too is possible. There are facsimiles of two German manuscripts in our collection which do somewhat resemble Phillipps, though the similarities did not seem as striking as in the French manuscript. Both are from Erfurt and date from the middle of the twelfth century—one precisely from 1147. The only parallel for the abbreviation ꝗ (*quod*) used in the Phillipps is found in one of these German manuscripts; in other manuscripts, qd seems to have been the accepted form of the abbreviation. If the Germanic influence in the runes is in fact well substantiated, then Germany is your most likely alternative. I am afraid that considerations of palaeography, given the small amount of study done on manuscripts of this period, can provide only broad guidelines for dating Phillipps, or for identifying its origin.

Not only the paleography but the literary style (if such it can be called) deserves scrutiny. Normally it is virtually impossible to place either geographically or chronologically a medieval manuscript on the basis of style alone. Not only was the grammar continually and rapidly changing, but there was a bewildering array of stylistic idiosyncracies due to general social and historical conditions, to the wealth, erudition, and even rationality of the establishments responsible for manuscripts and to the particular circumstances of the author or copyist. This is especially so in the case of a collection of recipes compiled over several centuries by no identifiable author, but by a series of unknown scribes. Yet some observations are viable in the case of our manuscripts.[5] *S* and *P*, although two hundred years apart, exhibit a relationship of tradition that is not shared by *L* and *K*. The former are, so to speak, closer to Classical Latin in vocabulary, grammar, and form. As noted elsewhere in this introduction, the cause of this is that they shared in the Classical revivals of the tenth and twelfth centuries. Thus their vocabulary, while containing many words borrowed from elsewhere, seldom shows traces of vulgarization as does *L*, e.g., *granci* (Italian *granchio*) for Latin *cancri*, or *K*, e.g. *indico* (Italian *indigo*) for *indicum*. In grammar the verb forms of the second personal singular present indicative, imperative, subjunctive and the future indicative are surprisingly interchangeable in all four MSS but far less so in *P* and *S* which tend to use the Classical imperative form for such orders as "Mix this," "grind this," "break this," etc., although even *P* has all four verb-forms in six lines in chapter 219. Form is more difficult to assess, but Hedfors notes in *L* many examples of confusion between e.g., the accusative and ablative cases, often occurring even when two words are in apposition with each other, which is not nearly so often the case in *P* and *S*. On the other hand some of these irregularities may be due to abbreviation, particularly in the case of final *m*, which is normally written as ~ over the preceding vowel in all four manuscripts, and a poor copyist, as apparently was the case in *L*, could easily write, for example *crocu* for *crocū* (=*crocum*). Spelling is irregular and internally inconsistent in all manuscripts but more so in *L* and *K*. Thus we find interchanges between *u*, *i*, and *y*, *ph* and *f*, *ch* and *c*, *c* and *g*, *f* and *b*, and *e* for *ae*. With regard to the last it appears to have been the general usage throughout the Middle Ages for the diphthong *ae* to be replaced by the vowel *e*; *P* and *L* do so regularly, though cases are not infrequently to be found in *S* and *K*. The Latin word for "and," *et*, is spelt out in *L*, but *K* and *S* use the ampersand (*S* using it also for the syllable *et* in other words) while *P* uses the more sophisticated sign 7.

On all these matters it is imperative to consult the discussion of word mutation, style, and grammar in the Lucca manuscript given by Hedfors (1932) and especially by Svennung (1941). A similarly detailed analysis of the *Mappa* manuscripts would be desirable: Here it is perhaps sufficient to repeat that

⁵ Hereafter the abbreviations *L, K, S,* and *P* will frequently be used for the Lucca, Klosterneuburg, Sélestat, and Phillipps-Corning manuscripts respectively.

stylistically *P* and *S* represent a higher, more classical tone and a more educated grasp of the traditional Latin language than either *L* or *K*. There is so little of *K* that it is hard to judge, although Ganzenmüller feels that it is poor in quality, and stands somewhere between the popularization or italianization of Lucca and the finer expertise of Sélestat and Phillipps. All of this suggests that the *Mappa* tradition is to be preferred to the Lucca, with the implication that northern Europe is more important than Italy for the compilation and preservation of medieval technological information.

Our conclusion based on the consideration of this paleographic and stylistic evidence combined with the surprisingly little that comes from the nature of the contents is that the *Mappae Clavicula* originated in northern France or contiguous areas of Germany. The slight presumption in favor of the former, mainly based upon the first definitely known location of the manuscript, is balanced by the hint of German pronunciation in the transcription of the runes (chapters 288-C and P) and the fact that it seems unlikely that a French writer would have designated as French the soap of chapter 288-D. There is no doubt that some of the material that was copied into this manuscript from a source other than Sélestat had been written in England and subject to a later German influence and, further, that the final manuscript is not English. Since items in both the fore and aft blocks of the additions to Sélestat in Phillipps had appeared in the earlier Klosterneuburg fragment, the manner of aggregation was obviously complex and cannot be established with any certainty.

The Lucca manuscript had its roots in a very ancient tradition. Sélestat grew from the same stock, but gathered in some important ninth-century additions. The Phillipps manuscript retains all of these and in addition contains material of probable English origin, some of it altered in a way to suggest Germanic influence. Altogether, the Phillipps-Corning Manuscript of the *Mappae Clavicula* stands at the very apex of the traditional compilation of recipes of chemical technology and it has a sampling of information on other topics.

SIR THOMAS PHILLIPPS'S PUBLICATION IN *ARCHAEOLOGIA*, 1847

Phillipps's first public announcement regarding his manuscript was in a form of a paper that was read on 22 January, 1846, before the Society of Antiquaries in London. The full text of the manuscript was published a year later in the Society's journal, *Archaeologia*, in which it occupied pages 183 to 244 of volume 32, published in 1847. It bears the cumbersome title "Letter from Sir Thomas Phillipps, Bart, F.R.S., F.S.A., addressed to Albert Way, Esq., Director, communicating a transcript of a MS. Treatise on the preparation of pigments, and on various processes of the Decorative Arts practised during the Middle Ages, written in the twelfth century, and entitled Mappae Clavicula." There is a certain justice in the fact that, following a misunderstanding by Berthelot, it is often listed in bibliographic notes under Way's name, for Way did far more than just transmit a communication.

The correspondence between Way and Phillipps on the editing of the *Mappa* is among the Phillipps's papers now preserved in the Bodleian Museum at Oxford, and photocopies were made available to us through the courtesy of Dr. R. W. Hunt, Keeper of Western Manuscripts. It is obvious that Albert Way (1805–1874) had a great deal to do with editing the transcript as well as persuading his Council at the Society of Antiquaries to authorize its printing in *Archaeologia* despite heated opposition by a radical group within the Society. The *Mappa* may have cost him his beloved directorship, for he was forced to resign shortly after. Moreover, although Phillipps saw the importance of his manuscript in connection with early painting techniques (then the subject of active research by Charles Eastlake, Robert Hendrie, Mary Merrifield, and others), it was Way who grasped its much broader significance. On 22 January, 1847, reporting that the last pages of his edited transcript were now in the hands of the printers, Way wrote to Phillipps:

. . . I enclose a proof of the letter, with which your communication of the Transcript was accompanied and brought before the Society this, according to customary usage should precede the text of the Clavicula, with the addition of any further observations, if you think proper to add any. Might I venture to suggest, as you commend it to notice *solely* on account of its interest in connexion with the Art of painting, that it is scarcely less valuable and attractive in regard to the curious information it conveys connected with metallurgy, the manufactures of glass, coloured leather, bone and horn, of soap, and with working ivory and crystal. The instructions for the formation of foundations of buildings and bridges, and the preparation of cements, are highly curious and valuable to the Architectural Antiquary, and not the least important portion of the Treatise is that which relates to the prototypes of Artillery, the flying arrows, and inextinguishable fires, respecting which most curious information is here to be obtained. Any one of these subjects would form an ample theme for a very interesting Dissertation, most properly and essentially subject matter for the Antiquaries—How mortifying then was the sneer with which my recommendation to the Council respecting Clavicula was met, that it might be more suited possibly for the Commission of Fine Arts. [Way to Phillipps, 22 January, 1847].

In this same letter Way calls attention to the existence and significance of the English words "goat tree" and "greningwort," referred to above. On the basis of this, Phillipps entirely rewrote the letter transmitting the paper to the Society and produced the version incorporating all these points which serves as preface to the published text. The original letter (undated but obviously written at about the time of the presentation at the Society's meeting on 22 January, 1846) consisted only of a brief description of the manuscript and its acquisition:

Mappae Clavicula
My dear Sir

This manuscript is a small 12mo Volume of 67 leaves on Vellum, written in the 12th Century, & was purchased at Paris in 1824 of the Reverend Mr Allard, Curé of St Eustache. It is entitled "Mappae Clavicula," i.e., The Little Key of Drawing. It appears to be quite perfect except a little cropping in one leaf. It has always appeared to me to be one of the most curious & interesting books relating to the Art of Painting in existence & under that impression I once begged The Royal Society of Litterature to print it in their Annual Transactions but they did not think it worth while [I would have printed it myself but the Booksellers have condemned all my Publications so that they cannot be sold in the market.] [6] It has therefore remained untill some Person of taste & judgement should think fit to give it to the World.

I am my dear Sir

                                        Very truly yours

Albert Way Esqr.   Thos Phillipps

Albert Way had previously worked on other manuscripts in Phillipps's collection, beginning at least as early as 1841, and the two men had corresponded frequently. Way's contribution to the publication of the *Mappae Clavicula* was far greater than is usually expected of an editor. He was one of the founders of the Archaeological Institute and it was during the first meeting of this organization, at Winchester in 1845, that the question of the publication of the *Mappa* seems to have first come up. The Institute sought help from the Society of Antiquaries to publish some of the surfeit of papers that had resulted from this meeting, and Way, who was an officer of both organizations, urged Phillipps to submit it for inclusion among these. However, on 18 December, 1845, after receiving the manuscript and seeing the magnitude of the editorial job before him, he wrote suggesting that it might form the first article in the volume scheduled to appear in the spring of 1847. It was Way

---

[6] The sentence in brackets is a contemporary insertion in Phillipps's hand. It is a reference to the publications of Phillipps's private press, the Middle Hill Press, which had indeed (though understandably) proved unpopular with the book trade. The publications of this press are overwhelmingly concerned with genealogy and English local history though they contain some items of technical interest. The first and by far the most important was a reprint in 1826 of Christopher Merrett's annotated translation of Neri's *Art of Glass* (London, 1662). Then followed a bit of science fiction entitled *A Fragment of the Voyages of Mr. Verigull Gulliver* (1832) which shows an interest in, but ignorance of, aeronautics, astronomy, and microscopy; and, oddly enough, a quarto 4-page phamphlet headed *H. Ste. Claire DEVILLE ON ALUMINIUM and its Chemical Combinations 1854*. This last is an abbreviated English translation of Deville's paper in *Comptes rendus* (1854: p. 279) together with other short contributions on the same subject by Chapelle, Wöhler, and especially Chernot. (Copies of the Gulliver and Deville pamphlets are among the Phillipps material in the Houghton Library, Harvard University.)

who presented Phillipps's letter describing the manuscript at one of the Society's meetings as a preliminary to getting Council approval for publication. When reporting this action to Phillipps on 27 March, 1846, Way remarked that the manuscript was not in a shape for the printer but he offered to prepare good copy himself. He wanted the scribe's abbreviations to be expanded because, he later said,

... the Mappa, as I would hope, may be taken up by practical men, such as Eastlake, or Hendrie, or Owen Jones, and be turned to useful account, as Theophilus has been already. The mere obsolete technical Latin of this kind of writing is already sufficiently discouraging—but to superadd the difficulty of reading contractions would be I think needlessly to make a discouragement to any use of the work, which would be very inexpedient. . . . [Way to Phillipps, 19 April, 1846].

In preparing the new transcript for the printer, Way worried about the same questions of style, paragraphing, and chapter enumeration that have bothered us 120 years later. In a letter of 30 December, 1846, Way plaintively refers to "the hopeless effort . . . of endeavouring to make a copy fit for the compositor without the opportunity of looking at the ms. which you unfortunately would not permit." Phillipps was a rather truculent individual, and his strange behavior in witholding the manuscript was only partly accounted for by the fact that he mislaid it for several months. He finally did offer to show it to Way in May, 1847, somewhat after he had returned the page proofs with his own corrections.

The correspondence also shows that first Mrs. Merrifield and later Sir Charles Eastlake (both of whom were famed for editions of early treaties on painting) were approached regarding the preparation of an English version of the *Mappae Clavicula* but it throws no light on the subsequent failure to carry through.

NOTE ON CHAPTER ENUMERATION

Only the first 66 chapters bear numbers in the original manuscript. The numbering of the chapters beyond this point was established by Phillipps and Way during the setting of type for the *Archaeologia* publication. The actual page proofs with Phillipps's and Way's corrections were preserved to be eventually included in Sotheby's Hodgson's-room sale on 15 May, 1969. They are presently in the possession of Messrs. Hofmann and Freeman Ltd. of Sevenoaks, Kent, who graciously permitted one of the translators to see them.

In a batch of page proofs of what are now chapters 61 through 192C (returned by Phillipps to the printers with the postmark 7 February, 1847) chapter 76 and 77 are unnumbered, so that 78 became 76. Some further omissions occur later, and, beginning with 117 (which is 113 in the proof) many chapters were marked

for subdivision into new ones with new numbers. In this proof the entire lack of numbering after chapter 177 was not amended. Even in a batch of later undated page proofs for the chapters following 205 there are still errors in numbering, and it is obvious that the numbers by which the chapters are now known were arrived at casually and late. The manuscript itself bears the published chapter numbers written in the appropriate places in the margin. They are in light pencil, in a hand that is neither Phillipps's nor Way's.

As indicated earlier, we have followed the original numbering of chapters, which has necessitated subdivision of many of the sections that were regarded by Phillipps and Way as a single chapter. Since it seems undesirable to change their widely quoted enumeration, we have kept it in this edition, but have indicated the manuscript's original division into chapters by alphabetic suffixes wherever necessary, e.g. Phillipps's chapter 176 has become 176, 176A, 176B, 176C, etc. through 176I. There is no subsidiary or other relationship implied by this designation.

In many cases when the Phillipps manuscript lacks a chapter title we have provided one from Sélestat and occasionally, when the contents require it, we have used a Sélestat title in preference to that in Phillipps. These are all noted by the appropriate symbol, *S* or *P*.

The final listing of chapters appears on pp. 23 to 26. On pp. 10 to 14 is given a table showing the location of the various sections in each of the three early manuscripts.

Of the 382 recipes in Phillipps, 100 had previously been in both Lucca and Sélestat, 109 had appeared previously only in Sélestat, 7 only in Lucca and 166 are unique to Phillipps. Of the last group, about two-thirds have a style and content that suggests an older source (similar to Lucca or earlier), while the remaining third though miscellaneous in character and appearing at the beginning, middle, and end of the manuscript read relatively freshly and may have originated not much before the time they were copied in by the twelfth-century scribe. There are eleven chapters in *S* that are not contained in *P*.

Phillipps contains two prologues, the first apparently relating only to the eleven unnumbered chapters on pigments, and the second preceding a listing of chapters and the main body of the text. This list differs considerably from that in Sélestat, though neither list conforms very closely to the contents of either. The concordance of chapters, given in the table pp. 10–14 reveals no apparent logic behind the reordering of the chapters in the different versions. The 209 entries in the list of chapters in *P* follow the actual chapter numberings in the text up to chapter 56, after which there are many interpolated chapters bearing numbers so that when the list ends on 209 it corresponds to Phillipps's chapter number 261. Following this chapter the actual text contains over fifty others that are not listed and most of these are not present in the older manuscripts.

TABLE OF CONCORDANCE BETWEEN MANUSCRIPTS [7]

Chapter no. this edition	Phillipps manuscript folio and line no.	Sélestat manuscript folio and line no.	Lucca manuscript folio and line no.	Chapter no. this edition	Phillipps manuscript folio and line no.	Sélestat manuscript folio and line no.	Lucca manuscript folio and line no.
Incipit	1.1	—	—	25	12v.13	11v.16	—
i	1.9	—	—	26	12v.19	12.1	—
ii	1v.1	—	—	27	12v.21	12.4	—
iii	1v.11	—	—	28	13.4	12.10	—
iv	1v.19	—	—	29	13.7	12.14	—
v	2.4	—	—				
vi	2.11	—	—	30	13.10	12.18	—
vii	2.19	—	—	31	13.13	12v.1	—
viii	2v.12	—	—	32	13.20	12v.9	—
ix	2v.18	—	—	33	13v.1	12v.11	—
x	3v.1	—	—	34	13v.12	13.2	—
xi	3v.19	—	—				
0	4.13	2.1	—	35	13v.17	13.8	—
1	7.13	6.1	—	36	14.11	13v.3	—
2	7v.8	6.15	—	37	14.14	13v.7	—
3	8.3	6v.14	—	38	14v.1	—	—
4	8.16	7.7	—	39	14v.15	—	—
5	8v.3	7.16	—	40	15.3	—	222v.29
6	8v.11	7v.2	—	41	15.9	—	—
7	9.16	8.5	—	42	15.13	—	—
8	9v.8	8.18	—	43	15.17	—	223.13
9	9v.18	8v.9	—	44	15v.2	—	—
10	10.5	8v.19	—	45	15v.5	—	—
11	10.12	9.5	—	46	15v.9	—	—
12	10v.8	9v.2	—	47	15v.13	—	—
13	10v.21	9v.17	—	48	15v.17	—	—
14	11.3	9v.19	—	49	15v.19	—	—
15	11.19	10.18	—	50	16.6	—	—
16	11v.5	10v.3	—	51	16.8	—	—
17	11v.11	10v.9	—	52	16.10	—	—
18	11v.14	10v.11	—	53	16.20	—	—
19	11v.17	10v.16	—	54	16v.6	—	—
20	11v.21	10v.21	—	55	16v.19	—	—
21	12.7	11.8	—	56	17.14	—	—
22	12.22	11v.2	—	57	17.21	—	—
23	12v.7	11v.10	—	58	17v.4	—	—
24	12v.9	11v.11	—	59	17v.8	—	—

[7] The main chapter numbers in the present edition are identical with those applied by Phillipps in his publication of the text in 1847, though many of his chapters have been subdivided and are here designated A, B, C, etc. The folio and line numbers cited correspond to the beginning of the chapter in question. The letter v following a folio number denotes verso; recto is unmarked.

The line identifications in the Lucca manuscript are based on the text as given by H. Hedfors, *Compositiones ad tingenda musiva* (Uppsala, 1932), those of Sélestat and Phillipps from photocopies of the manuscripts themselves, reproduced as Appendices A and B, respectively, in the present edition.

The first eleven chapters of the Phillipps manuscript (those preceding the 2nd *Incipit* and the list of chapters) were left unnumbered by him. We designate them by small Roman numerals, i to xi. They are sometimes referred to, following Hedfors, as A1, A2, . . . A11.

An asterisk (*) preceding a number in the *P* column denotes one of the few chapters that are also present in *K*, the ninth-century Klosterneuburg fragment described by Ganzenmüller

(1941). Folio 1 *recto* of *K* has chapters 167, 172, 173, 174, and 174-A; Folio 1 *verso*, chapters 175, 176, 176-A, 176-B, 176-C, 176-D, 176-E, 176-F, 176-G, 176-H, 176-I, and 177; Folio 2 *recto*, chapters 219 (the part on iron only), 99, 219-A, 219-B, 116, 220, 221; and Folio 2 *verso* contains chapters 221-A, 221-B, 221-C, 221-D, 221-E, 222, and 223. It will be noted that Folio 1 *recto* and *verso* contains mainly chapters that are lacking in *L* but present in *S*. The second folio covers recipes from the *L* tradition that appear also in both *S* and *P*, but they are given in an order that follows *L* more closely than either *S* or *P*.

The folio numbers in Lucca are related to the letter designations used by Hedfors and others as follows. MS fol. 217r is Hedfors A, fol. 217v is B, thereafter continuing as follows: 218 = C,D; 219 = E,F; 220 = G,H; 221 = I,K; 222 = L,M; 223 = N,O; 224 = P,Q; 225 = R,S; 226 = T,U; 227 = X,Y; 228 = Z,α; 229 = β,γ; 230 = δ,ε; 231r = ϛ.

For some *Mappae Clavicula* recipes incorporated in later versions of the quite separate works of Eraclius and Theophilus, as well as some others, see Johnson, 1939, pp. 102–108.

Chapter no.	P	S	L	Chapter no.	P	S	L
60	17v.14	—	—	98	24.13	—	—
61	18.1	—	—	99	*24v.3	—	228.18
62	18.5	—	—	100	24v.10	—	—
63	18.7	—	—	101	24v.17	14.1	—
64	18.9	—	—	102	25.10	14.16	211v.¶I & II
65	18.12	—	—	103	25.18	14v.5	211v.¶ III
66	18.13	—	—	104	25v.2	14v.10	223v.9
67	18.14	—	—	105	25v.4	14v.11	223v.10
68	18.17	—	—	106	25v.7	14v.14	223v.12
69	18v.2	—	—	107	25v.10	14v.19	223v.16
70	18v.19	—	—	107-A	25v.13	15.1	223v.18
71	19.12	—	—	108	25v.17	15.5 [8]	224.32
72	19.20	—	—	109	26v.8	16.12	—
73	19v.4	—	—	110	26v.10	15v.2	224v.17
74	19v.10	—	—	111	26v.15	15v.7	—
75	19v.17	—	—	112	27.5	15v.16	226.32
76	19v.18	—	—	113	27.14	17.6	226v.6
77	19v.20	—	—	113-A	27.16	17.8	226v.8
78	20.6	—	—	114	27.18	17.10	226v.11
79	20.9	—	—	115	27v.2	17.14	226v.16
80	20.14	—	—	116	27v.8	17.20	228v.20
81	20.17	—	—	117	27v.14	17v.5 ⎫ 213.5 ⎬	226v.22
82	20.19	—	—				
83	20v.1	—	—	118	27v.18	17v.9 ⎫ 213.8 ⎬	226v.26
84	20v.8	—	—				
				119	27v.19	17v.10 ⎫ 213.9 ⎬	226v.27
85	20v.13	—	—				
86	21.4	—	—	120	28.2	17v.14 ⎫ 213.13 ⎬	226v.32
87	21.19	—	—				
88	21v.1	—	—	121	28.3	17v.15 ⎫ 213.14 ⎬	226v.34
89	21v.4	—	—				
89-A	21v.6	—	—	122	28.6	17v.18 ⎫ 213.17 ⎬	227.2
89-B	21v.12	—	—				
89-C	21v.16	—	—	122-A	—	17v.20 ⎫ 213.19 ⎬	227.5
89-D	21v.17	—	—				
90	21v.20	—	—	122-B	—	18.1 ⎫ 213.20 ⎬	227.6
91	22.9	—	—	122-C	—	18.4 ⎫ 213v.1 ⎬	227.8
91-A	22.14	—	—				
92	22.21	—	—	122-D	—	18.6 ⎫ 213v.2 ⎬	227.11
92-A	22v.4	—	—				
92-B	22v.9	—	—	123	28.8	18.9	227.14
92-C	22v.10	—	—	124	28.15	18.14	225v.13
92-D	22v.14	—	—	125	28v.12	18v.11	225v.30
92-E	22v.17	—	—	126	29.5	19.5	226.8
93	22v.20	—	—				
93-A	23.4	—	—	127	29v.2	19.20	226.22
93-B	23.6	—	—	128	29v.7	19v.5	⎧226.27 ⎨ ⎩230.1
93-C	23.10	—	—	129	30.6	20.1	230.12
94	23.17	—	—	130	30.10	20.5	230.16
95	23v.7	—	—	130-A	—	20.7	230.18
95-A	23v.13	—	—				
95-B	23v.18	—	—	131	30.12	20.11	230.22
95-C	23v.21	—	—	132	30.16	20.15	230.26
96	24.4	—	—	132-A	30.20	20.18	230.29
96-A	24.7	—	—	133	30v.4	20v.2	230.35
97	24.10	—	—	134	30v.6	20v.3	230v.1

[8] In *S* chapter 108 continues to 15v.1, then to fol. 16 which is a small leaf inserted with the scribe's correction.

Chapter no.	P	S	L	Chapter no.	P	S	L
135	30v.15	20v.10	230v.10	174-A	*37v.15	48.19	—
136	31.11	21.4	230v.28	175	38.8	48v.11	225.31
137	31.17	21.10	230v.34	176	38.16	48v.18	225v.4
138	31v.3	21.14	231.6	176-A	38v.2	49.4	—
139	31v.7	22.7	227v.20	176-B	38v.5	49.7	—
140	31v.17	22.17	227v.32	176-C	38v.9	49.11	—
141	31v.20	22.19	227v.34	176-D	38v.10	49.12	—
142	32.4	22v.3	228.4	176-E	38v.9	49.13	—
143	32.14	22v.13	228.13	176-F	38v.11	49.14	—
144	32.20	25.7	217.30	176-G	38v.14	49.16	—
145	32v.6	25.13	217v.8	176-H	38v.16	49.18	—
146	32v.10	25.17	217v.12	176-I	38v.19	49v.2	—
146-A	32v.13	—	223.32	177	39.1	49v.5	—
146-B	32v.20	—	—	178	39.5	49v.8	—
146-C	33.9	—	—	179	39.8	49v.11	—
146-D	33.15	—	—	180	39.12	49v.14	—
146-E	33.21	—	—	181	39.14	49v.16	—
146-F	33v.9	32v.6	—	182	39.18	49v.19	—
146-G	33v.14	32v.11	—	183	39.20	50.2	—
146-H	34.1	32v.18	—	184	39v.3	50.6	225v.10
147	34.8	45.13	229.1	185	39v.6	50.9	223v.22
148	34.12	45.17	229.4	186	39v.9	50.12	—
149	34.15	45.20	229.7	187	39v.12	50.15	—
150	34v.1	45v.2	—	188	39v.15	50.18	—
151	34v.5	45v.4	—	189	39v.19	50v.2	—
152	34v.7	45v.6	—	190	40.5	—	—
153	34v.8	45v.8	217.3	191	40.16	—	—
154	34v.9	45v.10 / 212v.17	217.4	191-A	40v.1	—	—
155	34v.11	45v.12 / 212v.19	217.7	191-B	40v.18	50v.9	—
156	34v.13	45v.14 / 212v.21	217.9	192	41.8	50v.18	223v.25
157	34v.15	45v.15 / 213.1	217.10	192-A	41.10	51.1	223v.27
158	34v.16	45v.17 / 213.2	217.12	192-B	41.12	51.3	223v.28
159	34v.18	45v.19 / 213.3	217.14	192-C	41.13	51.4	223v.29
160	34v.19	45v.21	217.16	192-D	41v.2	28.10	220.6
161	34v.20	45v.22	217.18	193	42.19	29v.2	220v.12
162	35.1	45v.23	217.19	194	42v.13	39v.1	—
163	35.6	46.3	—	194-A	—	40.6 / 212v.1	—
164	35.8	46.5	—	195	43.16	—	—
165	35.11	46.7	—	196	43v.13	—	—
166	35.13	46.9	224v.21	197	43v.16	—	—
167	*35v.9	46v.3	225.3	198	44.1	—	—
167-A	36.16	47.6	—	199	44.4	—	—
167-B	36v.7	47.16	—	200	44.6	—	—
168	36v.20	47v.6	—	201	44.10	—	—
169	37.14	47v.18	—	202	44.19	—	—
170	37.16	47v.20	—	203	44v.3	—	—
171	37.17	48.2	—	204	44v.7	—	—
172	*37v.6	48.10	225.28	205	44v.13	—	—
173	*37v.10	48.15	—	206	44v.16	—	—
174	*37v.12	48.17	—	207	45.4	—	226v.16
				208	45.10	—	228v.20
				209	45.17	—	—
				210	45v.3	—	—
				211	45v.6	—	—
				212	45v.9	—	—

Chapter no.	P	S	L
213	45v.13	—	—
214	46.19	21.18	227.20
215	46v.1	21v.1	227.23
216	46v.20	21v.16	227v.9
217	47.4	22.1	227v.14
218	47.9	22.19	227v.34
219	*47.16	22v.19	228.18
219-A	*47v.9	23.21	228.31
219-B	*47v.13	23.18	228v.7
219-C	*47v.17	23v.3	228v.11
220	*48.6	23v.13	228v.27
221	*48.9	23v.16	229.13
221-A	*48.16	24.2	229.19
221-B	* —	24.4 / 213v.4	229.20
221-C	* —	24.7 / 213v.5	229.24
221-D	*48.17	24v.1	229v.3
221-E	*48.19	24v.3 / 213v.17	229v.6
222	*48v.6	24v.9 / 214.1	229v.14
223	*48v.14	24v.16	229v.21
224	49.1	25.3	217.25
224-A	—	25v.14	218.15
225	49.7	25.20	217v.34
226	49.19	25v.8	218.10
227	49v.6	25v.18	218.20
228	49v.11	26.3	218.25
229	50.1	26.9	218v.1
230	50.18	26v.1	—
230-A	50v.1	26v.4	—
231	50v.7	26v.8	218v.14
232	51.8	27.3	219.1
233	51.17	27.9	219.8
234	51v.4	—	219.31
235	51v.9	27.15	219.15
236	51v.17	27v.1	219.20
237	51v.20	27v.4	219.23
238	52.2	27v.6	219.25
239	52.9	27v.11	219v.3
240	52.17	27v.17	219v.8
241	52v.4	28.3	219v.14
242	52v.9	28.7	219v.16
243	52v.12	28.10	219v.19
243-A	—	28.13	219v.21
243-B	—	28.17	219v.25
244	52v.16	28v.4	220.1
245	53.1	29v.14	221v.16
245-A	53.13	30.2	221v.25
246	53.18	30.6	221v.32
246-A	—	—	222.15
247	53v.13	30.18	222v.10
248	54.5	30v.7	222v.24
249	54.9	30v.20	223.3
250	54v.2	31.10	223.19
251	54v.18	31v.10	—

Chapter no.	P	S	L
252	54v.21	—	222v.21
253	55.5	33v.3	—
254	55.19	33v.14	—
255	56.8	34.14	—
256	56.20	34v.1	—
257	56v.2	34v.3	—
258	56v.6	34v.6	—
259	56v.8	34v.8	—
260	56v.11	34v.10	—
261	56v.13	34v.12	—
262	56v.16	34v.14	—
263	56v.19	—	—
263-A	57.5	—	—
263-B	57.15	—	—
264	57v.4	41.1	—
265	57v.13	41.11	—
266	57v.16	41.14	—
267	58.15	41v.13	—
268	58.18	41v.17	—
269	58v.3	42.1	—
270	58v.15	42.12	—
271	59.9	42v.6	—
272	59.17	42v.19	—
273	59.20	43.1	—
274	59v.6	43.7	—
275	59v.11	43.11	—
276	59v.16	43.19	224.1
277	60v.18	44.16	—
278	61v.19	44v.23	—
278-A	62.5	45.3	—
279	62.10	45.10	224.29
280	62.13	—	—
281	62v.17	—	—
282	63.4	—	—
283	63.7	—	—
284	63.11	—	—
285	63.17	—	—
286	63v.9	—	—
287	63v.17	—	—
288	63v.21	—	—
288-A	64.1	—	—
288-B	64.3	—	—
288-C	64.9 (a)	—	—
288-D	64.21 (b)	—	—
288-E	64v.14	—	—
288-F	65.1	—	—
288-G	65.2	—	—
288-H	65.3	—	—
288-I	65.4	—	—
288-J	65.6	—	—
288-K	65.9	—	—
288-L	65.11	—	—
288-M	65.12	—	—
288-N	65.14	—	—

Chapter no.	P	S	L	Chapter no.	P	S	L
288-O	65.15	—	—	291	66.20	—	—
288-P	65v.1	—	—	292	66v.9	—	—
288-Q	65v.3	—	—	293	66v.19	—	—
289	65v.11	—	—	293-A	67.5	—	ǀ
290	66.3	—	—	294	67v.1	—	—

A NOTE ON THE TRANSLATION

This edition has for its principal aim the presentation in English of the technical information contained in the *Mappae Clavicula*. The translation was first prepared from the text of the Phillipps-Corning manuscript as the most complete, but it has been compared word for word against a photocopy of the Sélestat manuscript and Ganzenmüller's published text of the Klosterneuburg fragment. Discrepancies have been settled in favor of whichever seemed most philologically and technically convincing. Usually Sélestat gave the better reading. If doubtful points remained in those parts that were also in the Lucca manuscript, the two texts of the *Mappa* were compared with the latter text as given by Hedfors (1932). Wherever we have chosen a reading that differs significantly from the Phillipps manuscript we have indicated by a parenthetical [S] or [L] whether Sélestat or Lucca is followed.

We have been less concerned than previous scholars who have studied the manuscripts with the rather horrendous textual variations and have not annotated numerous differences in spelling and grammar that convey little difference in meaning. Arriving at the technical intent is a somewhat easier problem than that of philological criticism, for knowledge of the properties of matter often places a modern reader in a better position than a medieval copyist. Our translation is unfortunately but inevitably more intelligible than the original Latin in a good many places (a fact that anyone who has not seen the original will find hard to believe!) but our intent has always been to decide what was written in the Latin, preferably in an early uncorrupted form, rather than what should have been written.

The result may be safely used by readers of English to obtain a good idea of the content of this important compilation of medieval technological information. The serious scholar needs no warning of the dangers inherent in dependence on other peoples' interpretations, and in important cases will always revert to the original Latin, indeed to all the original Latin versions with their differences, and reach his own conclusions. Perhaps he should then proceed from the library to the museum laboratory in order to find out from the physical examination of extant objects what really had been done in medieval workshops.

A word is needed on our choice of title. The title of the work, *Mappae Clavicula*, is enigmatically redundant. Literally translated it is "A little key to the chart." It comes from the time when parchment *mappaemundi* displayed sailing directions to the whole known world; but why a key to a map? Phillipps wrote on the flyleaf of his manuscript "The Little Key of Drawing or Painting," and both the contents and Prologue do emphasize "painters' and other kinds of work." *Mappae Clavicula* might almost be taken to mean "Index of Abstracts," but the work itself is more nearly a compilation of compilations. Our translation, "A Little Key to the World of Medieval Techniques," is intended to catch the overtones of the two Latin words of the title and to reflect the contents more accurately than a more literal translation would have done.

THE *MAPPAE CLAVICULA* AS A SOURCE FOR THE HISTORY OF TECHNOLOGY

The *Mappae Clavicula* is second only to Theophilus's *De diversis artibus* as a written source for the study of medieval technology. It sits squarely in the main series of collections dealing with metals, pigments, and miscellaneous chemical operations that began with the clay tablets of Nineveh and Babylon (see R. C. Thompson, 1936), continued in Egypt with the Alexandrian Greek papyri now in Leyden and Stockholm, and in the ninth century passed both to Italy (Lucca codex 490, *Compositions Variae*) and to northern Europe with the first evidence of the *Mappae Clavicula* itself. There followed Eraclius, *De coloribus et Artibus Romanorum*, Theophilus, *De diversis artibus* (a completely original work far more realistically practical than the others) and excerpts of all of them continued to be recopied in innumerable different combinations to form eventually the basis of the many Books of Secrets that appeared shortly after the invention of printing and whose influence can even be seen in the household formularies of the present century.[1]

---

[1] The bibliography of the earliest printed versions is discussed by Ferguson (1888) and Darmstaedter (1926). The most complete and most popular was the *Secreti* of the pseudonymous Alessio Piemontese (Venice, 1555). In the eighteenth century the most important, even rather useful, ones were Godfrey Smith, *Laboratory or School of Arts* (London, 1740 and later); the anonymous French *Secrets concernant les arts et métiers* (4 v., Paris, 1724). This kind of literature merges indistinctly into

The earlier members of this series have long fascinated scholars. With literally hundreds of recipes mentioning such things as mosaics, purple, gold, glass, bronze, and a host of pigments, dyes, and changes of color, the chapter titles promise information on the background of medieval craftsmanship in metal and paint, on the very birth of alchemy, and on the beginning of modern science and technology.

Muratori (1739) who discovered the Lucca manuscript and first described it, Merrifield (1849), Giry (1878), Ilg (1874), Berthelot (1887a, 1887b, 1893), Berger (1912), Hedfors (1913), Burnam (1920), Johnson (1935, 1938, 1939), and others in the best tradition of medieval scholarship have examined the several manuscripts, and have traced their history and the influences of one or another school or region in the successive compilations. These show some influence of Dioscorides, Pliny, and other Greek and Roman writers, but in general they are quite independent of the mainstream of classical literature and make no pretense at literary cohesion. In origin they were certainly practical, however far from the workshop they later drifted. Superimposed on the direct continuity of some alloy recipes transcribed into Latin from the Greek sources, used in the Alexandrian papyri, there are later additions which originated variously under the influence of Italy, classical and Byzantine Greece, Egypt, the Middle East, and even Kent. Nevertheless none of the surviving manuscripts of the *Mappa* are the work of an author who knew the technical realities behind the words. The information was not even edited to produce a consistent picture, but was simply recopied with additions from other sources, occasionally more up-to-date, that seemed to relate to the general area. The compilers made no attempt to reconcile the many glaring discrepancies and duplications, and probably did not know enough even to notice them.

Two types of scholarship have been particularly fertile: philological studies based on word variants which reveal cultural contacts and influences, and art-historical studies in which mention of a given technique provides an indication of interest in it at the place where the compilation was made and occasionally provides a clue as to how various extant works of art were actually made. The former studies, of course, illustrate the truism that normally the earliest text is preferable, but in the case of the *Mappae Clavicula* there was hardly an original in the usual literary sense; the text was born of drudgery, not inspiration. Little attempt was made by any of the compilers to

the handbooks and dictionaries such as Dossie's *Handmaid to the Arts* (London, 1758), or Philippe Macquer's *Dictionnaire des Arts et Metiers* (Paris, 1761) and Andrew Ure's *Dictionary of Arts and Manufactures* (London, 1842), which were written to high editorial standards by experienced authors and today provide an excellent picture of accepted knowledge on various subjects even if they were not always at the very forefront of science.

bring together all the recipes dealing with a certain type of operation, or to rectify their inconsistencies. The additions are mainly in blocks of recipes without changing the previous grouping, although there are a few casual insertions of individual recipes. The several chapters in Phillipps which show obvious Arabic influence (196–201) are clearly a later interpolation, though they unaccountably appear in the middle of the main block of recipes from the Lucca tradition.

With ancient literary or historical works the primary aim of scholarship is to reconstruct an author's original words. This is impossible with manuscripts of the present sort, and interest has largely focused on the misreadings that arose because of the under- or over-zealousness of the scribes, their blindness and their misplaced ingenuity. A technological manuscript can throw much light on both the customs and customers of medieval scriptoria because the contents were not understood by the copyist and because his incorrect associations and his preference for one letter over another were less subject to the control of style or common sense than was usual in legal, literary, or ecclesiastical transcriptions.

Whatever may have motivated the original writers of individual chapters, the *Mappae Clavicula* as it now stands is a compilation of compilations. It was not edited and it shows no evidence of a critical intelligence or definable practical need governing the assembly. At every stage there would have been much other material that anyone really concerned with technical processes would have added, and certainly such a man would have rejected a lot that is included. The frequent unintelligibility and technical implausibility of the recipes in these manuscripts are not proof that they do not reflect the technology of the times in which they were written, although the mere longevity of any technical description is in itself a fairly good indication of its unreliability for this purpose. Moreover, compilations almost never represent current knowledge: those who write on current technology from their own knowledge usually ignore them. This means that the Lucca manuscript, the *Mappae Clavicula*, and others in their class are precarious sources, though they cannot, of course, be disregarded for at least they exist and they come from a period virtually devoid of other technological writing.

This mis-match between the written record and contemporary technology extends far back in history. In discussing the famous tablets on glass-making from the library of King Assurbanipal of Ninevah (668–627 B.C.), Oppenheim (1970) remarks, "In spite of their content, these texts cannot be taken simply as technical instructions. . . . They have to be considered . . . as literary creations within a complex literary tradition."

The *Mappae Clavicula* seems to have been preserved mainly because of the apparent connection of its recipes with the arts of the painter. There is nothing on ceramics and almost nothing on the most flourishing and characteristic arts of the day—architecture, sculpture, bronze-founding, and goldsmith's work. The compilers scarcely noticed such things, any more than they did the contemporary stirrings of a more powerful technology that would lead to vastly cheapened and enlarged production of everyday objects and to vast social changes. Though aesthetic motivation continued to lead to the discovery of technologically important materials and processes, the scale of technology was moving away from the artist's workshop and even more from the interests of those who wrote and preserved manuscripts. As with all beginnings, the most imaginative and portentous technology of the day was not appreciated enough to be recorded, at least in any writing that achieved the highly improbable state of survival to the present day.

Earle Caley (1926: p. 1164), in commenting on the incompleteness and redundancy of the recipes in Leyden X, states that they were "rather in the nature of reminders for skilled workers . . . than detailed descriptions for purposes of general information. . . . This papyrus was a kind of laboratory notebook of the operations of the chemical arts of the times." Possibly this was true of Leyden X (though that manuscript was in a form far too elegant for the workshop) but by the time similar recipes appear in Lucca and the *Mappae Clavicula* they are no more than a distorted echo of a much earlier period of technology. They comprise in the main information that would have been unnecessary for a practitioner and are given in a form that is quite inappropriate for the instruction of a novice. Perhaps they were once workshop receipts, but if so they reflect an earlier technology not a medieval one. Contemporary technology is, however, to be found among the later accretions, especially the initial chapters on pigments (i to xi), those on assaying and distillation, and some of those on incendiary stuffs, chrysography, and building. One cannot help but feel that the record of medieval techniques that is preserved in surviving objects is more valuable in checking and interpreting the written sources than the latter are for interpreting the objects.

This caution against over-valuation of the *Mappae Clavicula* and its relatives as sources for medieval technology does not deny them considerable significance, for there was a technological origin somewhere. The historian is constantly concerned with the degradation of his texts, and with the problem of the relation between what actually occurred or what was thought and what appears in the written record. The verbal poverty of the workshop aggravates relations with the scriptorium. However, the exclusion of poetic meaning from technological writings and the fact that they reflect a real world of physical substances and chemical reactions that was partly unknown to the scribes make them particularly useful

to reveal the transmission and decay of words and letters uninfluenced by concepts.

De Santillana and von Dechend (1969) suggest that the mnemonics associated with early astronomical observations became the basis of the common myths of mankind, which persisted long after they had ceased to be regarded as useful astronomical knowledge. There was no comparable poetry in verbalized chemical technology before mystical alchemy took it over, but perhaps the recipes were preserved by learned, or at least literate, librarians in the natural belief that what is written is more important than what is done. Moreover, doing so would assuage their bad consciences regarding their ignorance of practical things even if it was of questionable service to the practitioners themselves. Though the commercial motives which exist today for overproducing stuff of this kind were probably lacking in the Middle Ages, there were undoubtedly many custodians of manuscript collections who took a certain pride in being broad-minded enough to place technology beside theology, even if no one came from the local workshop to read it.

Today's scholar has an advantage over earlier readers of the *Mappae Clavicula* in that he can use modern knowledge of the properties of matter to identify some of the substances and processes. Thus, one can faintly glimpse through the verbal haze of chapter 69 the use of an annealing process to change the properties of glass. Both decadent Greek and chemistry confirm that *caucucecaumenon* is copper oxide, and *psimithin* is lead carbonate, but words such as *licamonia*, *anthimis*, *dedamia* and even such common words as *elidrium* and *lazur* occur in such diverse chemical contexts that identification with a single substance would be quite misleading. Many words that we have thought it safer to leave untranslated are undoubtedly corruptions of once meaningful words, if only in the jargon of the workshop.

In an article on the dangers inherent in translation, D. V. Thompson (1967) inveighs against the easy assumption of corruption in apparently meaningless phrases, rightly pointing out that these serve to flag unusual and therefore most interesting sections, and so call for particular effort at understanding without "interpretation." This is certainly true in connection with relatively literate sources like the earlier versions of Theophilus, but corruption nevertheless did occur. Many examples are provided by the changes that occur in the Lucca, Sélestat, and Phillipps manuscripts. For example, *aurocollon*, gold solder, which is utterly inappropriate to the making of an artificial stone in chapter 137 is a misreading by the scribe of *P* for *taurocollon*, ox-glue, as it appears in *S* and *L*. Similarly in chapter 220, *P* has the verb *tere*, grind, for the noun *terram*, earth, of *S* and *terra* of *L*, and in chapter 221-E the scribes of *P* and *S* refer to salt, *sale* and *salem*, where the lime, *calce* of *L* and *K*, is clearly correct. Fortunately, there is an alternative record of medieval technology. It is nothing but scholars' fondness for words that has led them to prefer the written record to the far better one preserved in medieval objects themselves. More can be learned from a detailed laboratory examination of medieval paintings, textiles, ceramics, and metalwork than could have been told by the makers themselves. As the importance of the technical part of man's past experience becomes more widely recognized it will be necessary to establish new kinds of libraries and reading devices, namely collections of artifacts associated with laboratory facilities to study them. Although a good start has been made on archaeological material from periods lacking verbal records of any kind, the extension of such studies into later periods has hardly begun. It will be noted that surviving examples of, for example, medieval metalwork indicate the use of a much smaller range of alloy compositions and techniques than would be inferred from the evidence of the *Mappae Clavicula*. In the manuscript, the on-going down-to-earth technology of the workshops is submerged in what seems to be a host of records of occasional experiments, local compromises, and trick solutions. In many parts of the *Mappa* wherein past scholars have seen alchemy, magic, or vital contemporary technique, the present translators see mainly corruption!

The above is simply a plea for the application of the historian's usual critical approach to sources of any kind. The Phillipps-Corning manuscript is the most complete exemplar of its genus, and it does contain several sections that were nearly contemporary additions and so reflect new technology. And, of course, from a period of which so little record exists even inadequate evidence must be examined for whatever it can yield.

What is the significance of the *Mappae Clavicula* for technological history? There is no doubt, as the great Berthelot and others have shown, that there was a continuous tradition from much earlier material joined with threads from many other sources. Johnson (1939: pp. 88, 89) who has analyzed the influences detectable in the Lucca manuscript, sums up as follows:

The *Compositiones Variae* is drawn from, or indebted to, the works of various peoples. This dependence has been traced back through many centuries by means of successively interdependent works. The collection is based mainly on Italian knowledge and practice, greatly influenced by Graeco-Byzantine learning and skill. The various works and peoples to which it is indebted are these: The Spanish influence is shown by palaeographical evidence, the parallels from Isidore, and the fact that extracts from his works are included at other places in the general body of the codex; its debt to the Arabs is manifested by its occasional use of Arabic words (this influence seeming to come by way of the Greeks); the Graeco-Byzantine source is evidenced by recipe no. 126 [*Mappae Clavicula* recipe no. 131], an obvious trans-

literation, drawn, with much probability, from a Greek source brought by an artist fleeing the Iconoclast persecutions; the parallels from Olympiodorus, Oribasius, and pseudo-Democritus, offer additional evidence of this important influence; recipe no. 67 [*Mappa* no. 43] is a direct word-for-word translation of recipe no. 73 [actually 74] of the Leyden Papyrus X, and manifestly attests a direct line of written tradition to third-century Egypt; the parallels from Dioscorides and Theophrastus show indebtedness to the early Greeks, especially when we remember the importance of the Leyden Papyrus, together with the fact that its last eleven sections are drawn direct from Dioscorides; the parallels from Pliny and Vitruvius and the recipes of Cato indicate that much of the same material was already in the hands of men not only in the early Roman Empire but even in the Republic; the quotations from the *Charaka* and the Assyrian Cuneiform Tablets are indications of indebtedness to the chemical knowledge (used in the industrial arts) accumulated by the ancient Hindus, Mesopotamians, and Egyptians (glass having been discovered in Egypt). We see, therefore, that our work, representing the knowledge of the arts accumulated up to that time, is successively indebted to Italian, Spanish, Arabic (via the Greeks), Graeco-Byzantine, Alexandrian, Roman, Greek, Hindu, Assyrian, and Egyptian sources. Like the Papyrus of Leyden, the *Compositiones Variae* utilized all available sources.

The compiler of the *Mappae Clavicula* incorporated all this: in addition he included some freshly translated accounts of Arabic alloys, north European runes, ancient Greek pneumatic toys, and a number of other recipes from various places on pigments and dyes, alcohol, sugar candy, coffer-dam construction, and incendiary mixtures.

Many of the ingredients of the recipes are identified with a place of origin, a very necessary precaution before the days of chemically purified substances. Vitriol, alum, steel, and even copper from different sources would be vastly different in composition and properties. Even today "Swedish" steel or "Straits" tin has some meaning. But it will be noted that almost all of the place designations in the *Mappae Clavicula* are classical in nature—an indication that the recipes are not contemporary with the compiling of the manuscript. None of the recommended simples come from the central or northern European sources on which the artisans in these areas must have depended in practice. The names would to some extent be retained as a kind of index of quality, though merchants would doubtless be willing to affix labels corresponding to any requested origin on to packages of whatever they had in stock.

The emphasis in the *Mappae Clavicula* is more on composition and admixture than it is on process. Some of the recipes convey straightforward information such as the composition of white copper, niello, and solders (the last including both alloys similar to today's hard and soft solders as well as the old reduction-and-alloying process of soldering using copper oxide on gold or silver), and the preparation of metallic pigments for the miniaturist. There are a number of recipes on gilding and others that seem to be aimed at giving a good color to gold by the superficial removal of alloy by chemical attack, while others aim at producing the superficial appearance of gold or silver on a base alloy.

The majority of the recipes are clearly intended to achieve some kind of decorative effect at little cost. The compilation almost certainly arose somewhere on the fringe of the decorative arts, made not by an artisan for his own use but perhaps to satisfy the curiosity of a patron of art who thought he might want to have someone reproduce wondrous effects of which he had heard.

The clearest recipes of all, both in the older sections and the new, deal with the production of pigments for the painter. It is rightly in connection with the history of painting that the manuscript has been most frequently studied in the past. In an excellent treatment of early medieval book illustration, Roosen-Runge (1967) gives the text of chapters vii, ix, x, and xi and many other paragraphs in the *Mappa* dealing with pigments, which he compares with similar sections in Theophilus, Eraclius, Le Begue (1431), and other sources that have been published. Roosen-Runge describes the results of microscopic examination of several illuminated manuscripts, mainly English, and reproduces in color over 380 fine low-magnification photomicrographs of various pigment combinations. His discussion of vehicles and pigments, being based on a critical study of textual material combined with microscopic studies of contemporary artists' work, is the most authoritative yet to be published.[2]

It should be noted that though there is a guildlike oath of secrecy in the Prologue, a cipher in the chapter on alcohol, and a reference to prayer in one of the now-missing chapters in the Sélestat list,[3] there is no hint of magic or of any desire to hide true meaning in symbolism of the sort that later beclouded alchemical literature. If many of the words are obscure, this is simply an opacity resulting from copyists' errors superimposed on whatever the first scribe set down in his record of the unfamiliar terminology used to describe strange operations on unheard-of substances.

The magic spell that Berthelot saw in chapter 288G is simply a misreading of a missing figure caption, and the startling mixture used in cutting rock crystal

(chapter 290) is, we believe, not a dependence on occult science as Thorndyke suggests but simply a result of the natural tendency, visible in many other chapters with less colorful materials, to build up and retain an unnecessary degree of complication in procedures that had once been found to work. It was magic of a kind, perhaps, because it involved forces not understood, but surely not the invocation of occult powers. When there were no purified chemicals in labeled bottles and no general theory to guide him, the artisan would not lightly change his practice. Moreover, the more spectacular recipes are the least likely to be omitted by a compiler: feeding a virgin goat with ivy and using his mixed blood and urine to carve crystal will impress the layman more than the suggestion simply to dip it in turpentine. There is a recipe for cutting crystal in Theophilus (III. 95) that is very similar to this one and he also has a quite wondrous recipe for Spanish gold (III. 48) but this attitude is even less typical of his treatise than it is of the *Mappa*. Both the magical overtones and the theoretical structure later imposed by the alchemists were, we believe, totally absent from the minds of the men who originated most of these recipes. The Theban mortar, Scythian vitriol, and Lycian or Arabian saffron were local supplies, not exotic ones, in the workshops where the recipes originated.

It certainly was not the intent of the *Mappae Clavicula* to provoke speculation. The man capable of seeing the broad metaphor of transformation hidden in all metallurgical operations is rarely the man who uses his hands and eyes to make the initial discoveries in the heat and clutter of the workshop. Operations of the type reflected in the *Mappa* with the truly marvelous qualitative changes in color and physical nature, did lead later alchemists (to whom words *per se* meant far more than they did to craftsmen or experimentalists) to regard such changes as a symbolic key to the nature of the universe. But this overlay of theory is totally lacking in the *Mappa* itself, both in the earlier texts and in the twelfth-century accretions. There is no hint of making gold, only of imitating it and making a limited amount of it go farther. Though the generation of stones in the various environments in the earth is frequently mentioned, neither astrological influences nor the Aristotelean elements are invoked. The *Mappae Clavicula* represents practice both unhelped by theory and unhindered by it, with each operation standing pretty much on its own feet. This attitude permitted the retention of a large number of recipes for achieving the same result, which today would require no more than a simple statement of the type of reaction involved. The recipes, however, though practical, must not be considered as representing the experimental method in today's meaning. There was undoubtedly a background of learning by trial and error—see, for example, the opening sentence in

---

[2] Roosen-Runge (1970) gives a summary of his views on the entire medieval color-recipe literature as part of a cooperative coffee-table book dealing with alchemy and its background. It was not available until after the present work had been completed.

[3] The title of the lost chapter (first noted by Berthelot [1893: pp. 55–56]) is on folio 5v, following a *capitula* entitled "A recipe for gold." This reads *Precem quam dicis quotiens conf. facis SS. autem [con]flas ut bonum exeat*—"The prayer which you offer, whenever you make a recipe is SS. And then you cast [the metal], so that it comes out good." Perhaps this is less an incantation than a simple request for a blessing at the beginning of the day's work. "SS" seems to have been explained in the next chapter following, *Interpretatio sermonum atque signorum*, "The Meaning of Speeches and Signs."

chapter 3—but there is no exhortation such as that in Biringuccio's *Pirotechnia* (1540) to "try everything, to vary the proportions and the means until the best way is found." Still less was there any attempt to design experiments to test a theory, or to deduce a theory from a critical experiment. There was no theory, no contact with philosophy, and it is rather surprising that the work achieved such popularity and longevity in the philosophically inclined Middle Ages.

The range of substances named is wide—chapters 192-C, 192-D, and 193 attempt a systematic listing of them—but the techniques and the apparatus involved are in most cases quite rudimentary. The most complicated device referred to is the glassmaker's furnace, which existed in both a small and a large form and with upper and lower units—undoubtedly the melting and annealing areas although there is a hint of a hearth for the preparation of frit, as in Theophilus. Lead plates with levigated emery are used to polish gemstones. Ingredients are mixed and ground in mortars or on marble slabs, are kept in earthenware pots (occasionally glazed), in brass ones, or, for dyeing, in bladders. There are open and sealed glass flasks. One of the commonest operations is *conflare*, to melt together ingredients, supposedly in a crucible in a forge fire with bellows. Distillation is inferred in the single recipe on alcohol (chapter 212) though the still is there described simply as "the vessels used for this business."

Operations on metals take up more than half of the manuscript, but they are rarely what might be called practical metallurgy. At best they are decorative techniques and are on the fringe of practicality even in that field. There is almost no hint of the magnificent character of medieval metallurgy as it is revealed in the preserved or excavated objects that are now in museums or ecclesiastical treasure rooms. Cloisonné and champlevé enamel are not mentioned, neither are the processes of raising, chasing, tracing and engraving involved in the making of chalices, nor the molds used in the casting of censers, aquamaniles, baptismal fonts, doors, and bells. There are several recipes for solders, but nothing on the making of filigree, beaded wire, or granulation. It is a chemist's view of metals, not that of a smelter or a smith. Metals are melted and alloyed, as well as treated with fluxes and other reactants. Except for gilding it, there is nothing whatever to do with iron. Recipes for compounding leaded bronze are given, but not a word on shaping it by casting. The hammer is used to beat metal into sheet or crush it to powder, never to forge it into a useful object or to drive a chisel or punch in decorating it.

Though gilding is mentioned, the current practical methods with leaf or with amalgam appear alongside the far older and costlier recipes involving the production of rather dilute alloys of gold and silver followed by treatment to produce superficial enrich-

ment of the precious metal. The written tradition of this technique seems to have vastly outlived its practical use in Europe. As we suggest in the note to chapter 2, this ability to produce a large bulk of alloy containing minor amounts of precious metal but with a surface composed of true gold or silver may have suggested to later alchemists the idea of a more subtle multiplication—a far jump from anything in the minds of the metalworkers who discovered the effect and knew its superficiality.

For the military engineer there are incendiary mixtures and tanklike battering rams in the *Mappae Clavicula*, but for his civil counterpart there is very little, and no mention of the contemporary successful attempts to harness the power of water and wind to do man's work. The main points of engineering interest are the building of a cofferdam for the construction of a bridge foundation under water and some information on the depth of foundation needed for various sizes of building in different soils (chapters 101, 102). In mechanism, there is the description of a universal joint of the type later called the Cardan suspension (chapter 288-0), and a brief listing of some pneumatic automata like those of Philo and Hero (chapters 288-G to -M). Almost everything else is chemical or metallurgical. In this the *Mappae Clavicula* reflects the curiously slow development of mechanical devices of an ingenuity comparable with that of the very early discoveries in metallurgy, ceramics, and dyeing.[4]

With all its faults or perhaps because of them, there are innumerable useful hints to be found on many aspects of materials, technology, and cultural influences. The table of runes with their Latin equivalents reflects an interest in northern Europe—but where? The multiplication table, so desirable with Roman numeration, gives a hint of subprofessional arithmetic, and quantitative analysis appears in the Archimedean assay method of chapter 194. Though alcohol was known somewhat earlier, the first reference to how it is made is in this twelfth-century manuscript. Incendiary materials are here described in far more detail than previously, some centuries before the classic *Liber Ignium* of Marcus Graecus. One of the oldest but least written-about professions, that of the dyer, contributes many recipes, with a curious emphasis on the dyeing of skins rather than textiles. The magnificence of medieval illumination could only come from the gold leaf, the ground gold inks, and the other pigments described in the *Mappa*, and the glory of medieval stained glass was based on recipes such as those given in chapters 256–258. The permanence of paintings depends on the use of linseed oil described for the first time (though in connection with gilding not as a painter's vehicle) in Sélestat, chapters 112–116.

---

[4] For a suggestion regarding the role of aesthetically motivated play affecting invention in these two fields, see Smith (1970).

Above all, the *Mappae Clavicula* serves to summarize and transmit the knowledge of the materials used by painters and other craftsmen that was part of the European heritage in the twelfth century A.D. If the compiler did not know enough to insert descriptions of the most advanced practice of his times, at least he did not intentionally delete the traces of an earlier experimental technology that we find so fascinating today. The following essay has very kindly been contributed by Jeffrey Hoffeld of the Cloisters Museum, Metropolitan Museum of Art, in order to place the art-historical value of the manuscript in clear perspective.

THE BACKGROUND AND PLACE OF THE *MAPPAE CLAVICULA* IN THE HISTORY OF ART[1]

Reichenau was an important center of both the artistic activity and the revival of learning stimulated by the Carolingian court. It is there that we find the first evidence of the *Mappae Clavicula*. The ninth-century manuscripts of the Reichenau schools, characterized by a sensitive integration of Merovingian and Carolingian forms, represent an important stage in the history of medieval painting. In fact, it is partially on the basis of Reichenau's achievements during the ninth century that many scholars have been led, perhaps mistakenly, to associate later Ottonian works, such as the Gero Codex, the Gospel Book of Otto III, and the Codex Egberti, with the continuing productivity of Reichenau scriptoria.

Reichenau is also among the important centers which revived the classics for their literary and linguistic value. In their efforts to revitalize and refine the Latin language, these centers, including Aachen (led by the learned Alcuin), Fulda, Corbie, St. Gall, and numerous smaller schools, fostered a *conservatio* as much as they brought about a *renovatio*. This, of course, was a natural by-product of their study of the past. The writings of late antique authors were thought to contain an as yet unspoiled Latin which could be used to reconstruct both the language and the culture of the empire before Charlemagne. The ancient texts became the reference section of the monastic library. They were copied directly and imitated; often the imitation was bound within the same covers as the model. This was the case with an old Lombard *Florilegium*, which came into the hands of the poet monks of Reichenau in the third decade of the ninth century. It contained excerpts from the writings of pagan and early Christian authors which were copied several times by the poets of Reichenau. One copy went to Laon where it was, in turn, copied. Another was sent to Mico of St. Riquier, author of

the *Exempla diversorum auctorum*, who revised it for teaching purposes. To the excerpts from Ovid, Statius, Vergil, Persius, Martial, and Juvenal, the works of Prudentius, Fortunatus, Theodulf, and Smaragdus of St. Mihiel were added.

In view, then, of Reichenau's position in art and learning, it is not surprising to find listed among the books in its library the *Mappae Clavicula de efficiendo auro volumen 1*. Although the manuscript is lost, this entry of 821–822 in the inventory of the monastery's library represents our earliest known mention of a *Mappae Clavicula* manuscript. This guide to materials and techniques, perhaps a copy of an older manuscript, was probably, like the *Florilegium*, used as a model for subsequent copies. The tenth- and twelfth-century versions of the Sélestat and Phillipps manuscripts of the *Mappae Clavicula* exist, no doubt, because of the care given in the ninth century to the preservation of existing examples such as that at Reichenau.

Like other antique writings housed in monastic libraries, the Reichenau *Mappae Clavicula* was probably a fountain into which the brethren dipped to purify their Latin. At the same time they found within its pages the curious and, by then, extremely remote techniques of their classical past. These recipes were often enhanced by the mention of a place name, indicating the supposed geographic origin of particular materials and methods—details which undoubtedly intrigued the reader. The book could not serve as a practical guide for the artist; perhaps much of it was not even intended as such but only to delight the uninformed reader with its description of changing colors and strange materials. While it could not train a beginning painter in the practical aspects of his craft, an artist would, nonetheless, value such a work, especially in Carolingian times. For it provides the atmosphere of the classical past; it reveals the mysteries of painting.

To understand this aspect of the manuscript's popularity, one must first consider several of the fundamental characteristics of pre-Carolingian art. Influenced by the arts of the migration period, the early medieval artist often stressed the inherent qualities of materials. In manuscript illumination, as in the fashioning of brooches and buckles, the play between the elements of design and the untreated or rough surfaces of the material is expressive. Imagery is affirmed, given substance and volume by the unpainted natural backdrop of an animal skin. In painting, there is little or no use of gold, the paint is often thin and translucent, revealing the skin beneath. The images themselves are often merely elaborate drawings, tinted.

Beside these pre-Carolingian practices, those of the Carolingian and Ottonian schools are drastically different. Like the few late antique manuscripts which exist, in the manuscripts coming from Aachen, for

[1] This section has been kindly contributed by Mr. Jeffrey M. Hoffeld, now Assistant Professor of Art History at the State University of New York at Purchase. He wishes to acknowledge the assistance of his student, Harriet Blitzer.

example, an attempt is made to transform the modest into the luxurious. Gold is extensively used, in lettering as well as in picture-making. The impression throughout is one of simulation. With paint, the qualities of marble columns, precious stones, and incised cameos are approximated. Many of the actual leaves of parchment are themselves transformed; stained purple, they are made to look like luxurious carpets on which jewels and objects in gold and silver have been placed around a modeled figure. The desire to embellish nearly everything with rich materials is suggested, too, in the text of the *Mappae Clavicula*. The reader is instructed in the painting of glass and in the staining of skins, in the naming of stones and in the making of gold leaf. These lessons, however fragmentary and imprecise, corresponded with the pursuits of the Carolingian and Ottonian artists. The richness of the late antique manuscripts in their possession was somewhat of a mystery to them; their luxuriance was startling. The explanation for these foreign qualities could be found, not in the specific details of the *Mappae Clavicula*, but in an attitude which it expressed: "*arte et ingenio vinci ingenium*" ("art and imagination subjugate the qualities of matter"). The manuscript became the voice of the past, taking the place of a long absent tutor, as an ancillary to the illustrated manuscripts and modelbooks of the late antique period.

Could its voice, however, be heard in the studio? There are well-defined relationships between the library and the atelier of a monastery which suggest that the two were only physically separate. The Rule of St. Benedict in the sixth century, as well as later monastic customs, prescribes a period of reading each day. While the minimal requirement was undoubtedly fulfilled for most by the reading of religious texts, one can imagine that some of the monks sought out their own interests among the books shelved in the library. An artist would have access to precious modelbooks and illustrated manuscripts which were probably stored, not in the shop where they could easily be damaged, but in the library. From the twelfth-century treatise of Theophilus, one has the impression that an interest in art may well begin in the library. Theophilus tells the reader that he has written his treatise *De diversis artibus* (*On Diverse Arts*) for "all those who wish to avoid and subdue sloth of mind and wandering of the spirit by useful occupation of the hands and delightful contemplation of new things."

Theophilus's treatise combines older recipes, which he probably found in the library of his own monastery, with what may well be teachings based on his own experience either as a metalworker or as an astute observer in a metalwork shop. It is clearly planned to instruct. The treatise as a whole not only serves to inspire the "delightful contemplation of new things" as does the *Mappae Clavicula*, but it also

provides a detailed practical guide for the use of men in the shop.

In the ninth century the intercourse between the shop and the library is even clearer. At Aachen, it seems there is complete interdependence of all components of the court: Charlemagne, the court and its entourage, the royal library, and the royal workshop. Einhard, the biographer of Charlemagne, embodies this remarkable synthesis. No stranger to libraries, having come to Aachen from the important monastic center of learning in Fulda, he himself was the head of a bronze-casting shop.

The twelfth-century manuscript of the *Mappae Clavicula* also suggests that the independence of the shop from the library should not be taken for granted. To its recipes, many of which are ancient, several up-to-date and useful directions have been added. These include the first twelve chapters on pigments, as well as chapters on incendiary stuffs, construction techniques, and chrysography. The inclusion of contemporary practices implies that the manuscript was of practical value if only as a mnemonic device.

The place of the Corning manuscript in the art of the twelfth century is similar to that occupied by the Sélestat and Reichenau manuscripts in the tenth and ninth centuries respectively. In the twelfth century, western Europe once again witnessed a revival of antique ideas and imagery. In jurisprudence, the study of the Latin language, poetry and literature, philosophy and science, and in art, the present was shaped in the image of the classical past. To some extent the twelfth-century revivals were predicated on the achievements of the Carolingian and Ottonian renascences. The classics studied in the twelfth-century monastic library were often those which had been copied or translated from earlier manuscripts by Carolingian scribes, editions of which were still available in the monastic libraries of the twelfth century. Abbot Suger's knowledge of the writings of the pseudo-Dionysius is the result of one such transmission from the ninth to the twelfth century. The twelfth-century *Mappae Clavicula* probably reflects a similar tradition, one which goes back at least to a tenth-century manuscript like the Sélestat. The desire to study, to record and to preserve the past is itself a sign of the renaissance movement. The author of the prologue to the *Mappae Clavicula* refers to such treatises as sacred books. For the twelfth century, they were sacred by virtue of their age, provenance, and the mysteries they revealed.

It is impossible for us to attribute individual works of art of the twelfth century to the influence of the practices recorded in the *Mappae Clavicula*. The manuscript probably served as a templet, rather than a source, against which the twelfth-century artist measured the richness of his own work, adding to it his own prescriptions for pigments and the fresh insights of Arabic sources.

For the student of medieval art, treatises like those of Theophilus and the *Mappae Clavicula* have a value which is far more profound than the immediate reward of identifying a particular recipe with the materials and techniques of known works. We labor under the assumption that in medieval art a familiarity with nature is manifested only in those attempts which appear to be naturalistic. We assume that only certain artistic forms, usually characterized by modeling and plasticity, reflect the presence of an awakened artist observant of nature. We forget that throughout the history of early medieval art there were no prepared packaged paints, inks, or parchment leaves. The locating of particular pigments required a familiarity with nature which was so intimate as to be incomprehensible to us today. It required a knowledge of not only the unchanging elements of nature, but of those that vary with climate, with geography, with the time of year. The eggs of a specific insect, at a specific time in its life, would yield a particular pigment. At other times the same eggs would be useless. The treatises reveal some of the difficulties. They take us along on the search for materials. They bring us into the shop and seat us beside the artist where we witness his every step. In Theophilus we can almost smell the furnace of the goldsmith's shop; in the *Mappae Clavicula* we marvel at the gold, glowing at the bottom of a pot. Finally, we may feel that this continual struggle with the resources of nature, preparatory to the fashioning of the work itself, leads, in the end, to a highly personal expression of nature, not as a physical phenomenon to be captured in realistic detail, but as an ungraspable force pictured symbolically and abstractly.

ACKNOWLEDGMENTS

This translation had its inception in a conversation between one of the translators and Paul Perrot, Director of the Corning Museum of Glass, at the time of the Xth International Congress of the History of Science in 1962. Mr. Perrot told of the museum's acquisition of the famed manuscript and offered its use. His continued interest and advice at all stages are deeply appreciated.

Photocopies of the relevant portions of the Sélestat manuscript were kindly made available by the Bibliothèque de la Ville de Sélestat through the courtesy of the Librarian, Dr. P. Adam, and the Centre National de la Recherche Scientifique.

Correspondence with the following gentlemen regarding various individual problems has been most helpful and is gratefully acknowledged:

P. Adam, Municipal Library, Sélestat
Robert H. Brill, Corning Museum of Glass
C. R. Dodwell, University of Manchester
Sidney Edelstein and Hector Borghetty, Dexter Chemical Co.
Eric P. Hamp, University of Chicago
Barry J. Hennessey, Harvard University
Martin Levy, State University of New York at Albany
John J. McCusker, University of Maryland
A. S. Melikian-Chirvani, Paris
A. N. L. Munby, King's College, Cambridge
Frank D. Prager, Abington, Pennsylvania
Heinz Roosen-Runge, Würzburg

The editors' indebtedness to Jeffrey Hoffeld of the Metropolitan Museum of Art will be shared by all readers of his important essay in the Introduction.

Finally, all authors must acknowledge the essential contribution made by their secretaries, but the work of Pauline Boucher and Linda Sayegh in carrying the translated *Mappae Clavicula* through many mutilated versions into visual if not conceptual clarity merits unusual gratitude.

The preparation of the first draft of the translation was aided by a grant from the National Science Foundation, its completion was made possible by funds for secretarial assistance from Grant No. H68-0-86 of the National Endowment for the Humanities. Our universities, the Massachusetts Institute of Technology and the University of Chicago provided the all-important environment in which the work was done.

THE MAPPAE CLAVICULA

LIST OF CHAPTERS IN THIS EDITION[1]

[1] See the footnote to the Prologue regarding the list of chapters in the original manuscript.

Here begins[2] *the book called* MAPPAE CLAVICULA

Every skill is slowly learned, step by step.
The first of the painter's skills is the preparation of
    pigments,
Then your mind should turn toward mixtures,
Then begin your work, but check everything by the
    fingernail
In order that what you have painted may be a thing
    of beauty and as freshly born.
Afterwards, as many talents have given testimony,
Skill will advance the work as this book will teach.

Chapter i. Vermilion

If you wish to make vermilion, take a glass flask
and coat the outside with clay. Then take one part
by weight of quicksilver and two of white or yellow
sulphur and set the flask on three or four stones.
Surround the flask with a charcoal fire, but a very
slow one, and then cover the flask with a tiny tile.

When you see that the smoke coming out of the
mouth of the flask is straw-colored, cover it; and
when yellow smoke comes out, cover it again; and
when you see red smoke, like vermilion, coming out,
then take away the fire, and you have excellent
vermilion in the flask.[3]

ii. Azure

If you wish to make the best azure, take a new pot
that has never been used for any work and set in it
sheets of the purest silver, as many as you want, and
then cover the pot and seal it. Set the pot in the must
that is discarded from a wine-press, and there cover
it well with the must and keep it well for 15 days.
Then uncover the pot and shake the efflorescence
that surrounds the sheets of silver into a shining bowl.
If you want to have more, repeat what has been
written above.[4]

iii. Again

If you wish to make a different azure, take a flask
of the purest copper and put lime into it halfway up,
and then fill it with very strong vinegar. Cover it and
seal it. Then put the flask in the earth or in some
other warm place and leave it there for one month.
Later uncover the flask. This azure is not as good
as the other, yet it is serviceable for painting on wood
and a plaster wall.

iv. Again

If you wish to make a third kind of azure, take
straw flowers, grind them and squeeze [the juice]
out of them into a very clean bowl. But first, on
wood and parchment, make the ground with white
lead; when it is dry, put the pigment on top and

---

[2] There is a second incipit introducing the prologue and a third
one at the head of the incomplete listing of chapters preceding
chapter 1. The first eleven chapters (here numbered i to xi)
are not numbered in manuscript *P* or in Phillipps's 1847 transcript, and they are entirely lacking in *S*. They deal exclusively
with pigments and are clearer than most of those that follow.
They are obviously a separately compiled group of contemporary
recipes prefixed by the twelfth-century scribe to the longer
manuscript.

[3] See also chapter 221-C. Gettens, Feller and Chase (1972)
discuss at length the history, identification, and use of vermilion
and cinnabar. In modern usage these terms are restricted,
respectively, to the artificial and natural forms of mercuric
sulfide (HgS). However, both *vermiculum* and *cinnabaris* are
used in different sections of the *Mappa* for what is always the
artificial product, and we maintain the difference in our translation.

[4] The chemistry of this process escapes us, unless the silver
contained enough copper to produce basic copper acetate by
reaction with the vinegar in the must, as in the next recipe. The
sealing of the pot would have to be imperfect to allow oxidation.
For a discussion of the various medieval blue pigments see
Roosen-Runge (1967) 2: pp. 51–59, also A. Raft (1968); D. V.
Thompson, Jr., (1936); and especially Joyce Plesters (1966). Raft
concludes on the basis of examinations of paintings of the period
that a rather poorly washed ultramarine made from lapis lazuli was
generally the *lazur* of the period. The soap-washing recipe described in our chapter 288 could perhaps have been used. Eraclius
mentions *lazur* in a context that would clearly distinguish lapis
from either azurite or vegetable blue. There can be little doubt
that all three types of blue pigment (and perhaps others) were
in actual use, to the confusion of those who study ancient techniques if not of the artists themselves.

continue making it in this way, until you see the pigment is like azure.

v. Green

If you wish to make Byzantine green, take a new pot and put sheets of the purest copper in it; then fill the pot with very strong vinegar, cover it, and seal it. Put the pot in some warm place, or in the earth, and leave it there for six months. Then uncover the pot and put what you find in it on a wooden board and leave it to dry in the sun.

vi. Again

If you wish to make Rouen green, take sheets of the purest copper and smear them all over with the best soap. Put the sheets into a new pot and then fill it with very strong vinegar. Cover it, seal it, and put it in a warm place for 15 days. Then uncover the pot, shake the sheets over a wooden board and put [the product] in the sun to dry.

vii. [Ceruse and] minium

If you wish to make minium, either red or white, take a new pot and put lead sheets in it, fill the pot with very strong vinegar, cover it and seal it. Put the pot in a warm place and leave it there for one month. Later take the pot, uncover it, and shake out whatever surrounds the lead sheets into another earthenware pot and then set it on the fire. Stir the pigment continuously and when you see the pigment become white, like snow, take away as much as you like of it, and that pigment is called ceruse.[5] Then put the rest on the fire and stir continuously, until it becomes red, like other minium. Then take it away from the fire and leave it in the pot to cool.

viii. Various Pigments

Pigments that are thick and clear on parchment[6] are the following: azure, vermilion, dragonsblood, carmine, folium, orpiment, byzantine green, *gravetum*, indigo, brown, saffron, red or white minium,[7] the best black from vine charcoal. All these pigments are tempered with glair.

---

[5] Ceruse is here *cerussa*, as also in chapters 63 and 94. In chapters iv and vii it is *album plumbum*, white lead, but elsewhere in the *Mappa* it is referred to by the latinized Greek word psimithin, which we have also translated white lead. The process involved here is the formation of basic lead acetate, which converts to the carbonate by reaction with atmospheric CO_2. (See note by Earle Caley in his translation of Theophrastus [1956: pp. 186–191]). The chemical similarity between the making of ceruse and verdigris (chapters v and vii) will be noted. Heating ceruse under oxidizing conditions produces first litharge (PbO) and then minium (red lead, Pb_3O_4), with invisible stages of deepening colour in between. The process is an old one; another good description of making it is in Theophilus (I-37).

[6] I.e., brightly colored pigments of good covering power for use in manuscript illumination.

[7] See Roosen-Runge, 1967: 2: p. 20.

ix. Mixtures [of Pigments]

If you want to know the nature and mixtures of these pigments and which ones are incompatible with each other, listen carefully.

Mix azure with white lead, darken it with indigo, lighten it with white lead.[8] Darken pure vermilion with brown or dragonsblood, lighten with orpiment. Mix vermilion with white lead, and make the pigment that is called rose; darken with vermilion, lighten with white lead. Again, make a pigment with dragonsblood and orpiment; darken with brown, lighten with orpiment. Darken carmine with brown, lighten with red minium. Make a second rose pigment from carmine and white lead; darken with carmine, lighten with white lead. Darken folium with brown, lighten with white lead. Again, mix folium with white lead; darken with folium, lighten with white lead. Darken orpiment with vermilion, and there is no lightening for this, since it turns all other pigments dung-colored.

Now if you want to make *Gladus* green, mix orpiment with black; darken with black, lighten with orpiment.

If you want to do similarly, take azure and mix it with white lead; darken with azure, lighten with white lead; and when it is dry, cover it with clear saffron.

x. Tempering [Pigments]

Temper Byzantine green with vinegar; darken with black, lighten with the white that is made from [the ash of] stag horn. Again, mix green with white lead; darken with green, lighten with white lead. Darken *gravetum* with green, lighten it with white lead. Darken saffron with vermilion, lighten it with white lead. Darken indigo with black, lighten it with azure. Again, mix indigo with white, darken it with azure, lighten it with white lead. Darken brown with black, lighten it with red minium. Again, make rose from brown and white lead; darken it with brown, lighten it with white lead. Again, mix saffron with white lead, darken it with saffron, lighten it with white lead. Darken red minium with brown, lighten it with white lead. Again, mix minium with brown; darken it with black, lighten it with red lead. Again, make flesh-color pigment with red lead and white; darken it with vermilion, lighten it with white lead.

xi. Pigments that conflict with each other

Now if you want to know which pigments conflict with each other,[9] here it is. Orpiment is not com-

---

[8] The lightening [*matizare*] and darkening [*incidere*] refer to the tempering of a basic color to represent the areas of light and shade on an object.

[9] The incompatibility is, of course, due to chemical reaction—the formation of black lead sulphide from cinnabar and ceruse, or the changing of pH of folium. Eraclius has a fine paragraph on this, reading, in Merrifield's translation, thus "*On colours in-*

patible with folium, green, red lead or white lead. Green is not compatible with folium.

If you want to make grounds, make a beautiful rose from vermilion and white. Again, make a ground from folium tempered with lime. Again, make a ground from green tempered with vinegar. Again, make a ground from green itself, and when it is dry, cover it with cabbage [juice].

If you want to write in gold, take powdered gold and temper it with glue made from the same parchment as that on which you are to write, and write with the gold and glue close to the fire, and when the letter is dry, burnish it with a very smooth stone or a boar's tooth. Again, if you want to make a robe or another painting, as I said above, put the gold on the parchment, darken with ink, or indigo, and lighten with orpiment.

Here begins the PROLOGUE of the following work

Since I possess many wonderful books written on these matters I became anxious to produce a commentary, not that I may appear to be encroaching upon the sacred books and [therefore] despite much labor accomplishing nothing, but that, avoiding that mortal heresy, I will disclose to those who wish to understand these things what the actual processes are that are used in all painting and other kinds of work. I call the title of this compilation *Mappae Clavicula*, so that everyone who lays hands on it and often tries it out will think that a kind of key is contained in it. For just as access to [the contents of] locked houses is impossible without a key, though it is easy for those who are inside, so also, without this commen-

tary, all that appears in the sacred writings will give the reader a feeling of exclusion and darkness. I swear further by the great God who has disclosed these things, to hand this book down to no one except to my son, when he has first judged his character and decided whether he can have a pious and just feeling about these things and can keep them secure.[10] Now although we have many other things worth saying concerning the virtues of what is written, we will begin the list of chapters.[11]

End of Prologue

Here begins the Book called MAPPAE CLAVICULA

1. Making the most gold

Take 8 oz.[12] of quicksilver, 4 oz. of gold filings, 5 oz. of fine silver filings, 5 oz. of Cyprian copper filings, 2 oz. of brass filings, 12 oz. of cleavable alum and the efflorescence of copper which the Greeks call *calcantum*, [blue vitriol[13]], 6 oz. of gold-colored orpiment, 12 oz.

---

compatible with each other.—Now, if you wish to know which are the colours that are incompatible with each other, they are these: —Orpiment does not agree with folium, or with green, or with minium. Nor does green agree with folium, namely, in the mixture of the materials of the said pigments, and in the works in which they are employed together. And these discordances are not in the mere [optical] qualities of the pigments, nor in their accidents of colour; for there are no colours, or qualities of colours, either simple or mixed, which, as regards the colour only, do not agree with any other sorts of colours in mixtures, namely, for composing other different mixtures; and you may thus have at pleasure almost innumerable varieties of colours. But the said discordances are, and are to be understood as being, in the other natural conditions, incident to the substance of the said pigments, they being contrary to each other in such manner that, if they are mixed together, one substance, by a certain natural incompatibility, either changes the other or is changed by it; and so the quality and beauty of the pigments themselves, as well separate as mixed, and their own substance, and the work done with them, are spoiled and destroyed. They therefore do not bear to be mixed together; and so, in the art of painting, besides the consideration that is to be had for the varieties of colour, and these and other things relating to the said art, we must not forget the proper and necessary considerations, drawn from a true theoretical and practical knowledge of and acquaintance with the natural conditions and contrarieties existing in the materials and liquors of the said colours, and of the contrarieties of the other things incident to that art." (Merrifield, 1849: 1: pp. 252-254.)

[10] For other allusions to secrecy see chaps. 13, 14, and 212.

[11] In the manuscript *P* this is followed by a list of 210 chapter headings which we do not translate. It covers a majority of chapters Nos. 1 to 261 of the present series in virtually the same order but with abbreviated titles and, because of omissions, usually with different numbers. Supposedly, like the Prologue just included, it once related to a previous shorter version of the *Mappae Clavicula* (much like the Sélestat manuscript of the tenth century though including two of the chapters showing Arabic influence) which had been added to and somewhat corrupted before the present version was compiled. The chapters not mentioned in the list are our chapters i to xi, 57 to 59, 62,65 to 69, 71, 72, 74, 91, 95, 114, 139, 146-A to 146-D, 156, 158, 161, 172 to 182, 184 to 194, 197, 198, 212, 213, 219-A, 219-B, 221, 222, 229, 238, 239, 245-A, 249, 251, 252 and 260. The Sélestat manuscript begins with a different unnumbered list of 196 chapters, naming many that are not actually included.

[12] In instructions for weighing out ingredients, both *L* and *S* commonly use either the sign for ounce (÷), or the word *uncia*. The sign ℔ is used (supposedly for *libra*, pound) for lb., in some of the earlier chapters in *S*, though *P* uses either ℥ or ℥, the standard symbols for dram or ounce, in the same recipes. Elsewhere in *P*, however, units have frequently but not invariably been changed to drams, though often ÷ remains and sometimes z or ℥ appear, all meaning ounce. To confuse matters further, Phillipps in his 1847 published transcription often reads ÷ and sometimes ⫶ for ÷. Other units such as *libra*, *siliqua*, *denarius*, *obol*, *cyathus*, *tremis*, *mina* and *aurei*, are spelled out or conventionally abbreviated. The symbol ð for pennyweight comes twice in *P*, chapters 276 and 277, and once in *L*, chapter 43. The intent was, of course, to give the relative fractions of the various ingredients rather than absolute weights, and in many cases they are given simply as parts. In the translation we have generally followed *S*'s preference for ounces in the chapters that are in that manuscript; otherwise we follow *P* without annotation. Anyone who tries out the recipes must, like any good cook, use his discretion!

[13] *Calcantum* (*chalcantum*). The terminology of the various natural and artificial crystalline sulfates and other salts that resembled them is extremely confused. The original Latin terms conveyed a far less precise identification of the chemical substances than do the modern English equivalents that we have perforce supplied. Confusion is especially great in manuscripts such as the present, for sometimes the words are simply survivals

of *elidrium*. Then mix all the filings with the quick-silver and make it like a wax salve. Put in the

of ancient recipes from the arid eastern Mediterranean world and beyond, while others represent later northern surrogates that were frequently chemically different though going by the same name. The Greek word *chalcanthos*, which literally means flower of copper as stated here, may be verdigris but it was most probably the soluble sulfate. Both Pliny and Dioscorides say that the best variety of *calcantum* was blue, which makes it copper sulphate (and indeed copper does seem to be necessary in those *Mappa* recipes that call for it), yet, *chalcanthos* was also related to *atramentum* which did not necessarily contain copper, any more than copperas does today.

In order to maintain the verbal distinctions of the original we have always translated *calcantum* as "blue vitriol"; *atramentum* as "vitriol," *vitriolum* as "vitriol [*vitriolum*]"; *misy* as "misy"; and *misi ciprii* as "Cyprian misy," obviously copper sulfate. However, the chemical distinction would by no means consistently follow the terminological differences. These substances would all be mildly corrosive soluble hydrated sulfates of iron, copper, or perhaps occasionally zinc—that is, variants of present-day copiapite, chalcanthite, or goslarite and the artificial substances related to them. Of all these salts, misy, usually a ferric sulfate such as copiapite, was the most strongly acid and corrosive (see footnote to chapter 3). *Altramentum*, which is frequently mentioned in the *Mappa*, is certainly a form of vitriol, probably green vitriol, ferrous sulphate. In classical Latin and even in some of Eraclius' recipes this word refers to a black pigment, but not here.

Other mineral salts that appear in the *Mappa* are as follows:

nitrum, natron, an alkali of uncertain composition, originally the efflorescence on desert rocks consisting mostly of sodium carbonate and sulphate. It is not to be confused with nitre or saltpeter, the natural efflorescence on walls adjacent to decomposing organic matter and which cannot be uniquely identified before the thirteenth century (Partington, 1960). The Latin *natronum* appears only in chapters 195 & 201 in recipes derived from the Arabic, where we translate as "soda" to avoid confusion with the vaguer earlier term.

alumen, alum, probably most often the astringent material that goes by that name today, potassium alum. *Alumen rotundum*, *alumen scissile*, which we translate literally as rounded and fissile alum respectively, are perhaps other sulphates that crystallize easily from solution in good polyhedra.

afronitrum (a transliteration of Greek *aphronitron*, foam of nitrum) is some kind of alkali, formed as an efflorescence on natural deposits of natron or as a scum rejected during some preparative processes. See chapter 192-D. In chapter 280, *afronitrum* comes from soap-making; in 221 it is a compounded flux for soldering. Bailey (1929: p. 170) suggests that it may be natron that has been heated to make it more caustic.

sal gemma or *sal montanum*, rock salt, is one of the few clearly identifiable substances.

sal ammoniacum, sal ammoniac. By the twelfth century this was probably the modern material. In earlier usage *hammoniacum* (which appears in chapters 72 and 124) was an organic substance, "The gum of Hammon."

tartarum, tartar, potassium tartrate, the refined deposit from wine fermentation. Burnt tartar, i.e., potassium carbonate, is referred to chapter 202.

Elidrium is believed by Svennung (1941) and most earlier writers to be the same as *chelidonium*, swallow-wort, a word which often appears in the *Mappa*. Although a plant such as this would not be inappropriate in some of the recipes (notably chapters 46, 60, and 68) there are others in which *elidrium* seems to be a brittle yellow mineral substance. Orpiment is usually one of its associated simples. In chapters 83 and 209 it seems to be a kind of electrum-like alloy. Swallow-wort would not answer all these diverse uses, and we have chosen to leave *elidrium* untranslated.

elidrium and orpiment at the same time; then add the efflorescence of copper and the alum and put it all in a pan on the coals and cook it lightly, sprinkling over it with your hand an infusion of saffron in vinegar and a little natron and sprinkle 4 oz. of saffron bit by bit until it dissolves and leave it to intermix. Then when the mixture has coagulated, take it off and you will have gold with increase. Now add also to the above ingredients a little moon-earth, which in Greek is called *Aphroselenos* [i.e., foliated selenite].

2. Again, making gold

Melt 1 oz. of silver, ½ oz. of copper and 1 oz. of gold. Again, take sand and press it onto a level place. Cool until it is dry, and mix again some salt and roast in the furnace for a day and a night. Afterwards take it out and wash it until the salt runs off; and again dry it and knead it in vinegar and set it aside for a little until it absorbs it and dries out. Then again put into the furnace a piece that has not been washed and do this once and again; knead it in vinegar every time you put it into the furnace. Now you ought to put it into the furnace four or five times until it becomes almost as if it is all cooked away; and when you take it out use a silver withdrawing tool, which in Greek is called *elquison*. Take a weight equal to the former amount, mix it all together and grind it.

Then melt separately the two kinds of material that you have concocted [i.e., the residual gold and the salt-containing cement which now contains silver] and sprinkle them gradually [onto a lead bath (?)] until it is used up. Then cool it and you will find that hard lead has been made Melt this together with *cepsonium*, i e., kneaded ashes. As is shown according to the Key, *psomion* is ash kneaded with water, which you lay underneath in the furnace to the thickness of a finger.[14]

3. Again

Take only a little for experiment when you do it once until you learn it thoroughly. Take 1 oz. of reddish Cyprian copper in solution [?, *posios*], 1 oz. of quite good silver, and melt it with chaff, until when hammered out it does not make a noise, and then melt it together with 1 oz. of gold and the same

[14] This chapter is a confused account of the cementation of a gold-silver-copper alloy by prolonged heating in a sand-salt-vinegar mixture, followed by alloying the residual metal with lead (supposedly in a scorifier) and finally cupelling on a bed of washed and compacted ashes to yield gold. The cement is separately scorified and cupelled to recover the silver absorbed in it. It does not, of course, make gold as promised by the title, or even increase its apparent amount as would the recipes in the next chapter and others in which chemical treatment of the surface of gold alloys is used to improve their color. For a description of the furnace for cementation see chapter 246-A.

amount of natron. Then turn face to face two little bowls, i.e., two hollow earthenware pots, and put inside them the widened (i.e., hammered out) melt that has been prepared, and mix in *antisma*. What had been a little bead of copper is now turned into a little bead of 4 oz. of silver [*S*]. In the bead we find more than an equal amount of gold. [Take] one part of Pontic sinopia, 2 parts of common salt, grind them all together, lay the sheets on the bottom and sprinkle [the sinopia and salt mixture] over them and coat them with pot clay so that they cannot breathe. Put fire under them until you feel it is all right. Take them out and you will have the very best gold.[15]

4. *Again*

Mix 4 parts of silver, 4 parts of Cyprian misy, 7 parts of pounded and sifted *elidrium* and 4 parts of sandarac. Also melt silver and sprinkle the above materials on it, and melt with a strong fire, stirring everything together until you see the color of gold. Take it out and quench it in cold water and keep this

---

[15] This chapter provides a moderately clear description of a process for giving a pure gold surface to an alloy containing approximately equal parts of copper, silver, and gold. The sheet metal would acquire a silvery-white coating as a result of annealing in air and pickling (as in blanched billon coinage) but the final cementation in sinopia and salt would eventually remove everything but gold from the surface.

Other recipes incorporate misy with the salt and sinopia. Misy, ferric sulphate, is as potent as sulphuric acid and it acts, even when cold, to corrode the alloy deeply, leaving *in situ* a layer of nearly pure gold that is porous but easily consolidated by burnishing or slight heating. This mixture is actually more effective than a cementation in brickdust and salt, for heat sinters the porous alloy and so inhibits penetration of the corrodants at the same time that it enhances the continual diffusion of the underlying base metals to the depleted surface. The chemistry of the reaction with misy is closely analogous to the assayer's parting operation with nitric acid acting on inquarted gold (an intentionally-made alloy of gold diluted with three times its weight of silver). The decorative use of parting by nitric acid may perhaps have preceded its quantitative application, and it is interesting to note that sulphuric acid, the successor to misy, returned to dominate the commercial separation of gold and silver in the nineteenth century.

Related processes appear in chapters 4, 16, 19, 20, and 27. Chapter 26 is headed "A doubling of gold." Although this recipe does not specify the necessary dilution by alloying, the adjacent ones do, and there is little doubt that the operation was the basis of the later alchemists' favored process known as multiplication.

The production of a pure gold surface by this kind of low-temperature reaction on a dilute alloy of gold was one of the most important techniques used by South American goldsmiths in the millennium before the Spanish conquest (see H. Lechtman, 1973), and it was also used by the Japanese in gilding their fine Oban coins. In Europe it was replaced by more economical methods of gilding with amalgam or gold leaf, although gentle heating with vitriol, salt, and iron oxide continues to this day as a means of very superficially improving the color of legal karat gold alloys.

For a general discussion of the role of color in the history of metals see Smith (1974).

mixed concoction in a bowl. Equal parts of Cyprian misy, sandarac and *elidrium*: and make a soft unguent of them, melt the silver, and while it is still hot, pour it into the same unguent.

Nota Bene[16]

5. *The recipe for the most gold*

Take copper that has been hammered out when hot, and grind filings of it in water with 2 parts of crude orpiment so that it becomes as viscous as glue, and roast it in a small pot for 6 hours and it will turn black. Take it out and wash it off, then add an equal portion of salt and grind them together. Then roast it in the pot and watch what happens: If it is to be white, mix in silver, if yellow, mix in gold in equal portions, and it will cause wonder.

6. *A recipe for gold*

Take 2 parts of goat's gall, 1 of ox gall and 3 times as much *elidrium* as of the above materials, and grind them with vinegar for 10 days. [This is the first composition.] Then take Lycian or Arabic saffron and grind it in a Theban mortar in the sun during the dog days; and add very sharp vinegar and grind it until the saffron disappears and is used up. Now put in a sixth of vinegar, the more the better, and let it dry. Then take the frog-green ore which others call copper, grind it very fine, and mix in saffron. [This is the second composition.]

Then take {i.e., gold *crisantimum*, which painters use, which they call *pusa*, [*lampusa*, *S*] and grind it in the same way, and you will have it for use. [This is the third composition.]

Now use it in the following way. Take as much silver as you want and melt it and add finely ground salt, and stir it continuously until the silver melts.

Afterwards let it coagulate, and while it is still hot pour it into sea water. Then melt it again and into 1 pound of silver put 2 oz. of the first composition, stir it vigorously and again let it coagulate. Then put it again [into the fire], melt it and add 1 oz. of the second composition. Stir it very well in the same way, and again pour it out. Again, melt it a third time and add a sixth of the third composition, and you will have it. Now, stir with a new iron rod. Watch it whenever you stir it, and when it seems to have become uniform, again mix in silver until it appears to you to be good.

These are the recipes: First this one.

Take 2 parts of goat's gall, 1 part of ox gall, 3 times as much *elidrium* as of the above materials, i.e., 9 parts, and grind them with vinegar for 10 days.

---

[16] Written in the margin in a later hand.

7. The second recipe

Grind Lycian or Arabic saffron in a Theban mortar with very sharp vinegar in the sun during the dog days until the saffron disappears and is used up. Put in a sixth part of vinegar, and the more the better and let it dry. Then take frog-green [ore] which is copper, and grind it very fine, and mix in saffron. The third recipe: Take *crisantimum*, which painters use, which they also call *pusa*, and grind it in the same way.

Now, you will have these compositions ready and according to what has been said above, you put into 1 pound of silver 2 ounces; into an ounce of silver, 2 scruples of the first composition. Of the second mixture or composition, into 1 pound of silver put 1 oz., into an ounce of silver, 1 scruple. Again of the third composition, into 1 pound of silver you put 1 ounce, into an ounce of silver, 3 siliquae.

8. A recipe for gold

You hammer out several sheets of pure silver and lay them down onto a preparation, which will be revealed, and sprinkle on them [more of the same preparation] and melt them until they are all reduced into one. Now, this preparation is what is called *Offa*: Take 4 scruples of gold, 1 oz. of Macedonian glue, 1 oz. of live sulphur, 1 oz. of natron, an entire pig's gall, [S], 2 oz. of soot [S], 1 oz. of Spanish minium, a whole fox gall, 1 oz. of *elidrium*, 1 oz. of Lycian saffron, and you make a potion containing iron in which you put all this together, and with this preparation [do] as above, and you apply it underneath the sheets and sprinkle it on top. Now into 1 pound of silver you put 1 oz. of the preparation alum [S] melt, and it will be gold.

9. Again

Take as much gold as you want and twice as much imported misy and the same amount as the misy of thick *pyriapian* filings of good copper or Cyprian copper that has been melted. Mix them both together and make a gold tube into which you deposit the three preparations. And so melt it, cooling as necessary, and take it out of the furnace and wash it off. You will then find the greatest weight of gold; and when this has touched the fire, it will become better. Mix in some cadmia or Trachian stone—it is yellow and laminar, i.e., flat or milky [S]—until it appears to you perfect gold.

10. Again, a recipe for gold

Take 2 parts of pyrite, i.e., firestone, and 1 part of good lead and melt them together until they become like water. After this add lead in the furnace until they are well mixed. Next, take the product out and grind 3 parts of it and with it grind 1 part of good *calcitis* and roast them until it turns yellow. Melt some copper which you have previously cleaned and add some of the preparation to it by eye, i.e., to the best of your judgment. It will become gold.

11. Again, excess of gold

Take $\frac{3}{4}$ of the purest gold, $\frac{1}{4}$ Cyprian copper, [and $\frac{1}{4}$ *magnesia* (S)] and melt them together. File it with a goldsmith's file and add 7 oz. of quicksilver, grind them together, and add a little vinegar and a little salt until the quicksilver absorbs the filings, and it will form an amalgam. Leave it to dry for 7 days in a glass jar. Afterwards take 4 siliquae of prepared native sulphur, some sandarac prepared from the *pusca salsa*, and 2 siliquae of yellow *bubulus*, and some orpiment that is made from Scythian vitriol,[17] and 1 siliqua of vulture gall. Grind them together and spread them over [other] amalgam. Carefully stop up the mouth of the jar and lute it with gypsum and roast it in the upper compartment of the furnace for three days and nights; and on the fourth day transfer the jar to the lower compartment of the furnace, so that it may become a sort of straw color. Then take it out and lay it aside. Now take 3 *cyathi* of silver and a quarter of certified gold and melt them and you will find out how it behaves, a sacred and praiseworthy secret.

12. Again, making gold

Take 5 *cyathi* of gold, 2 *cyathi* of certified copper and melt them together; then file them finely and add 16 quarters of quicksilver that is made from minium[18] and grind the filings together. Add a little very sharp vinegar and some salt, until the quicksilver absorbs the filings and it will become an amalgam. Then let it cook for 7 days. Now this is the preparation: 1 siliqua of sulphur, 1 siliqua of sandarac, 1 siliqua of orpiment made from Scythian vitriol, and 1 siliqua of vulture gall. You grind all these together and spread out the amalgam in a flask. Lute the mouth of the flask with gypsum and roast it in the upper compartment of the furnace until it becomes straw colored. Now take 4 quarters of the silver called *signatum* [i.e., stamped by an assay office?] and 4 quarters of the gold that is in the flask. Melt them together and you will find it.

13. The coloring of gold from horny copper which should disappear

Grind 1 part of copper, 1 part of ox gall, 1 part of roasted misy, heat them and you will see what happens.

---

[17] *Scithicum atramentum.* See footnote 13 to chapter 1. It is hard to see how orpiment could be made from it.

[18] The confusion between minium (red lead oxide) and cinnabar (red mercury sulphide) was not uncommon.

14. The coloring of gold which does not fail

Take 1 oz. of fissile orpiment, 4 oz. of pure, reddish sandarac, 4 oz. of the substance of magnesia, 1 oz. of Scythian vitriol, and 6 oz. of Greek natron, like ordinary natron. Grind the orpiment extremely fine until it is like mud. Now mix them all together and add very sharp Egyptian vinegar and ox gall. Grind them together and make them muddy and dry in the sun for 3 days. Grind and put in a flask and roast in the furnace that you know about for 5 days. Afterwards, take it out and grind it, add 5 oz. of ground gum, then add water and make it like mud and you will form a salve. Now you take 1 part of prime gold and 1 part of the salve. Melt the gold and put it into the salve. When the gold becomes green and so that it can be ground, melt 1 part of the colored gold and 1 of silver and you will find gold. If you want to make it prime gold, melt 4 parts of the colored gold and 1 part of common gold, and you will find the best and proven gold. Keep this as a sacred thing, a secret not to be transmitted to any one, and you will not as a prophet have given it away.

15. Another recipe for gold

Melt together 4 parts of copper and 1 part of silver and add 4 parts of unburnt orpiment, i.e., raw eureos [S] and after heating it very strongly, allow it to cool and put it in a pan. Coat it with potter's clay and roast it until it becomes cherry-red. Take it out, melt it and you will find silver. And if you roast it a great deal, it becomes elidrium. And if you add 1 part of gold to this, it becomes the best gold.[19]

16. Again, the same

Take gold made in this way, hammer it out into sheets the thickness of a fingernail and take 1 part of Egyptian sinopia and 2 parts of salt. Mix them together and apply a first coat of it over a sheet and when you have coated it over, seal it with potter's clay and roast for 3 hours. Then take it away and you will find the best gold, without fault.[20]

17. Making green gold with or without melting it

1 part of liquid alum, 1 part of Canopian balsam (which is used by goldsmiths), 2 parts of gold; melt these all together and see what happens.[21]

18. Making proven gold

Two parts of armenium, 1 part of zonitidos [S]. Grind them all, add a fourth part of bull's dung and an equal part of cadmia. Melt it and it will be rather heavy. Do the same thing also in copper.

19. The cooking of gold

Mix equal parts of ground salt and fissile alum into 1 oz. 3 scruples of Cyprian misy and grind them together. Prepare a new fire-pan; make thin sheets of gold and stack them one on top of another [with the compound] between. Fill up with charcoal and set it on fire three times.[22]

20. (A recipe for) gold

2 oz. of iron rust, 2 oz. of loadstone, 2 oz. of foreign alum, 7 oz. of murra stone and some gold: grind them with wine; it is extremely useful. Now there are some people who do not believe what a great usefulness there is in humors—these are the people who do not themselves make a demonstration. But if they do this, they will have to grant that there are some things to be marveled at. However, they should do it this way: i.e., by mixing them, melting them together and putting them in a goldsmith's furnace. When the bellows are applied, its nature will be found.

21. Making a heavier gold

When you work in perfecting gold, in order that your labor and care may not be in vain but will also produce some profit, the making of a mixture will be no mean advantage. When the gold has taken on the appearance of a fiery color, take 2 pennyweight of filings of Indian iron, grind them very well and put them into the gold. Then hammer out a sheet [from the gold], but not a thick one. Now, use misy and hide it in another [sheet]; and when it seems to be going well, heat it a little, so that it seems just to have touched the hearth. Then when

---

[19] The same recipe reappears under different titles as chapters 83 and 209. In the former, elidrium becomes electrum in the text, but not in the heading. The word euri in S is omitted in P, chap. 15, while in 83 it becomes curatum and in 209 eurem. Pliny (37, para. 161) has the word eureos (genitive euri) "an unknown precious stone" (dict.). The normal accusative form would be eureum, easily corrupted into eurem. The curatum may then be explained as a scribe's misunderstanding of eureum (c and e being readily confused) and his attempt to emend from an unfamiliar word, eur/e/um, to a familiar one, cur/at/um. At all events it clearly refers to unburnt orpiment. In some recipes orpiment is used in the burnt condition (supposedly then As_2O_3), in others it is specified as golden-colored, while it is most often used raw. The operation described in this chapter would make a white copper-silver-arsenic alloy which would hardly answer to any realistic test for a yellow metal (except perhaps superficially if its gold content were enriched by oxidation and pickling as described in footnote 15 to chapter 3) but it might make a fair fake electrum. In most other places where it appears, elidrium seems to be a mineral, for it is brittle enough to be crushed and is yellow.

[20] Chapter 246-A describes a method similar to this for superficially cementing sheets of a gold-silver alloy. See also chapter 3.

[21] Berthelot (1893: 1: p. 38) points out that this is a corruption of an older recipe to test gold. Recipe 43 in Leiden papyrus X also describes a procedure to detect the alloy in gold by the colors formed by superficial oxidation.

[22] $\bar{\imath}$. An abbreviation for ter.

you take it away, put it in a new pan and cover it. Now, in the cooking of a sheet, use misy. Clean it energetically and weigh out as much as you want. Depending on how much there is, mix in 2 parts of natron and 1 part of black lead.[23] Mix the unguent with water; arrange it on leaves or on a thin piece of cork and, when you have dried it, melt it with black lead and leave them until they are burned into each other and there remains what you want.

22. *The uniting of gold and its reshaping*

Take 5 oz. of gold and make tubes and take equal amounts of brass filings, fissile alum, Cyprian misy, and rock salt. Melt until everything is separated. When you have hammered out the preparations, shake them out from there, put a tube into the melting furnace, and some Theban natron, and so melt it. When you weigh it out, you will find it has been doubled. When this is put into the fire and beaten, it will recover its original color.

23. *Again, the melting of gold*

Melt 2 parts of gold, 2 of silver, 1 part of copper sheet.

24. *How an assay of gold should be made*

4 parts of silver filings and one part each of cadmia, sinopia made from misy, and burnt copper: grind all this together and wash with wine [S]. When the mixture is clean, make a little pellet [of it]. The nugget will melt and join into one. With this melt 4 parts of gold.[24]

25. *An operation with gold*

Melt 2 parts of gold, 2 of silver [S] and 1 part of copper sheet. Mix 4 parts of silver filings with the above-mentioned materials and grind them all together with wine and wash them. When this mixture has been made clean, make a little pellet and it will be in the form of a nugget. When this has been melted add to it 1 part of gold and so melt it all together.

26. *A doubling of gold*

Prepare 4 oz. of gold, 2 oz. of misy and 2 oz. of sinopia. Melt the gold until it is lively. Mix in those two (the misy and sinopia) during the melting, and take it out.

27. *Another way*

1 part of gold, likewise 1 part of silver and 1 of copper: make a sheet the thickness of a fingernail.

Coat this below and on top and add 1 part of roasted misy as a coloring matter. Roast for 2 hours; take it out and you will find the gold doubled.

28. *Another way*

Prepare 1 part of prime brass filings, so that they can easily be melted, 8 *minas* of Samian cadmia, 8 parts of roasted misy, 12 of *demnas* [S]. Melt carefully with this mixture.

29. *Another way*

Mix 1 [part of] the juice of the carpathum tree, which is a teardrop like the gum from the tree in which *arborinum* is engendered. Though some people want the herb or tree [that grows] in Egypt.

30. [*A gold ink*][25]

Grind together minium, mountain sand, gold filings, and alum with vinegar; and cook it in a copper pot, and stir. The color of this writing lasts for years.

31. [*A golden sealing wax*]

Mix 2 oz. of reddish natron and 3 oz. of minium. Grind with vinegar, add a little alum and leave it to dry. Then grind it and lay it aside. Take about half an obol of gold filings and 1 oz. of gold-colored orpiment. Mix them all together, grind them and pour over them pure gum soaked in water. Take it out and seal what you want, whether a letter or tablets. Leave it for two days and the seal becomes hard.

32. *Another way*

Put in juniper juice, which is called *bero inbriome.*

33. *Making hard gold fluid, so that it comes out of the fire better*

File refined gold with a fine file and thoroughly comminute it. Put it in a mortar of serpentine or rough porphyry, where it can be ground well. Add very sharp vinegar and grind them together. Drench it as long as it is black and pour it off. But when the vinegar gets its own color, then at last you put in either a grain of salt or at least *afronitrum* and it is made liquid so that one can write with it. Then you store it, as in a kind of mine, in a large glass jar, with

---

[23] Follows *S, et plumbi molibdinis. Molybdinum* was loosely used for plumbago (black lead) and molybdenite as well as for some compounds actually containing lead.

[24] This is a recipe for an alloy and has nothing to do with the assaying promised by the title.

[25] The titles of chaps. 30 and 31 actually both read *aliter,* "another way." Chaps. 30 to 35 and 68 to 71 describe the preparation of gold paints or inks for chrysography. In order that the gold should disintegrate and not compact together by grinding it is necessary to mix it with salt, honey or other material that is later washed away, or to render it brittle by alloying with mercury or lead, which is subsequently removed. Many of the inks were "extended" with yellow materials such as orpiment, litharge, or saffron. Sometimes no gold at all is used. All need some adhesive. See Alexander (1966).

a little gum to hold it fast. In this way also silver, brass, and iron can be made liquid. Now, in order to make the gold shine after you have written with it, rub the letters with the shell of a sea snail or with a boar's tooth.

34. *Liquefying gold*

Take 1 part of honey [*melaneum, S—elan*], 3 parts of orpiment, and 1 part of gold. But thoroughly hammer out the gold as much as you can, and cut it up with shears. Add 1 part of quicksilver and 1 part of very sharp vinegar and mix them all together. Grind them thoroughly in a mortar and add gum, and write with this ink and burnish it with a tooth.

35. *Making liquid gold*

2 parts of gold-colored orpiment, 1 part of *elidrium* and 1 part of litharge, whose color should be golden.[26] When you have ground these, pour them into a pot. Next take 24 square gold sheets. Grind as much as you want of these in a clean pharmacist's mortar, with the addition of a little salt. When it looks to you like ground sand, add fresh water, grind and wash it off; but do this by continually adding water until you see that the gold is pure. Then add those preparations [ground orpiment and litharge and] a little ground gum, but not enough to make it sticky. Drip onto it an extract of saffron and grind everything together, so that it has the consistency of ink. Place it in a shell or a glass pot; and when you want to use it, first smear the reed pen with liquid alum, then dip it in the gold and write. When it is dry, rub it thoroughly with a tooth.

36. *Softening gold so that you may form a seal in it*

Cook down some fissile alum in water, put in quicksilver and grind them together in a mortar, sprinkling in 1 part of ocher and saffron with pure glue and calf gall. Grind them together and use them.

37. *The liquefying of gold*

Take tin, melt it with quicksilver, and leave it to cool. Grind it thoroughly in a mortar with fissile alum and cover it with the urine of a boy. In this way it will become liquid and when it is of the consistency of a scribe's ink, write your work with it. When the letters are dry, separately grind Lycian saffron with pure glue and with this write [over] what you had already written, and when it dries, rub it with a tooth. But if the tin solidifies, [i.e. compacts during the grinding] melt it again and mix it well with [more] quicksilver.

38. *How gold can be made liquid without fire*

Take thin gold and silver sheets; grind them in a very hard mortar with salt and Greek natron until they are indistinguishable and seem to you to be thoroughly ground. Then you add more water in the same way and wash it off; and, when the gold remaining in the mortar is pure, grind in a little efflorescence of copper and ox gall, and so write with it. Keep it in a glass pot and when it has become dry grind it so that it becomes more lively. Now write with a reed pen or with a painter's brush. If you want what you are writing to be spread more lavishly, separately grind 4 parts of fissile orpiment and 1 part of *elidrium*, together with the above-mentioned substances, sift and mix. Grind as much as seems to you to be equal to the gold in the mortar. Grind, as described above, and write. When it is dry, polish it with a woolen polishing cloth, or with natron. With this you also paint on glass, on marble and on statues.

39. *Writing gold letters*

Take a sheet of malleable gold, cut it up into tiny pieces and put it into a glass pot. Add enough quicksilver, leave it until the gold liquefies, and transfer it to a mortar. Grind the quicksilver thoroughly. Now, when you see that it is well ground, add misy and copper, and grind the same way until the quicksilver is no longer distinguishable. Then put in misy until it looks to you like copper, and leave it to dry. Add sufficient liquid glue, grind and, using a paint brush, write whatever you want to have in color in the way that you have learned.

40. *Another way*

You melt lead several times and quench it in cold water. Then melt some gold, quench it in the above-mentioned water that has been used for the lead, and it becomes brittle.[27] Then you grind the gold thoroughly with quicksilver. Thoroughly clean the dregs as you know how, mix in liquid gum and write after you have dipped the reed in liquid alum. Clean the alum with salt and the best [S] vinegar.

41. *Another way*

Dye gold [powder] with Indian dragonsblood and put it in a glass pot. Surround it with coals, and it will immediately liquefy and become fluid enough for you to be able to write with it.

---

[26] Although litharge is naturally yellow, a pigment bright enough to be called golden might perhaps be what is now known as tin-lead yellow. This has not been identified in any actual painting, however, until about 1300 A.D.

[27] Although gold containing lead is indeed brittle and easily reduced to powder consisting of individual crystal grains, it is hard to see how molten gold could pick up enough lead from the water to be affected. The idea, however, persisted for centuries. *Cf.* chap. 71 which seems to be an independent Latin translation of an ancestor of this chapter in some other language.

42. Another way

Take tin and rub it with your fingers; and when they begin to get black, rub gold with them until it takes on the same blackness. Then melt it, solidify it, grind it and do as you have learned.

43. Making inscriptions with gold, according to the first method

1 part of *elidrium*, 1 part of broken resin, the white of 5 eggs by number, 1 part of gum, 1 part of gold-colored orpiment, 1 part of tortoise gall, 1 part of *carcale* scrapings. The weight of them all, after they have been pounded, should be about 20 denarii.[28] Then add 2 drams of saffron. This works, not only on papyrus and leaves of vellum, but also on marble and on glass.

44. Another recipe for gold

Mix native sulphur, the skin of a pomegranate, the insides of figs, a little fissile alum, and liquid gum. After adding a little saffron, write.

45. Again, another recipe

The yolks of 3 eggs, and the white of 1, and 4 oz. of gum, and 1 oz. saffron and 1 oz. of rock-crystal filings, and 7 oz. of gold-colored orpiment. Grind all these thoroughly together. Dry them for 2 days on end and then they are put back into saffron. So you write what you want.

46. Another writing in gold

1 part of *elidrium*, 1 part of orpiment, 1 part of tortoise gall, 1 of fissile alum and 1 part of the skin of a pomegranate which is gold-colored inside, 1 of gum, and 5 eggs. So let the weight of all these be 20 oz. [Add] 2 drams of saffron.

47. Another golden writing without using gold

Cook down in a pot the juice of a mulberry or fig tree and [add] to the juice a quarter part of alum, and smear it on the vessel that is to be gilded and so gild it. Before doing this smear it with a quarter [part of] myrrh, or the juice of a mulberry or fig tree, and so gild.

48. [Another kind]

Melt red natron and salt together, and smear it with water and you may make whatever work you want.

49. [Another kind]

Take a lump of [caked powdered] gold, put it in a glass pot and add ox gall. When it has been thor-

oughly thickened, leave it for 3 days. Then when you have learned that it has become liquid, slowly pour out the gall, and add salt water to the liquid gold. Now, transfer both to a clean copper pot, and warm it. Take it out and dilute it. Then, when it is dry, mix in liquid glue and so paint. Certainly, if it should not become liquid in 3 days, leave it alone.

50. Gold-colored writing on parchment, marble and glass, so that it seems to be made of gold

Mix gold with verdigris, grind it, and smear it on. Or grind quicksilver with a woman's feces, and smear it on.

51. The gilding of glasses, on a reed, and on copper

Put hyacinth color, which painters use, into salt and stir until it becomes liquid. Then smear the gold with it and dye it, up to four times.

52. Getting the gold color that you want

40 oz. of gold-colored orpiment, 15 drams of quicksilver, 10 oz. of chrysocolla, 20 oz. of foreign wine, 1 of *elquima*, 2 obols of black lead, 2 oz. of sulphur, 20 oz. of Gallic copper filings, 4 oz. of very white cucumber leaf: Pound these all together and sift them in a small coarse sieve and throw away the woody stuff that remains. Then knead them with the white of 50 eggs and let them dry. Pound them again and knead them with liquid gum and eggs, until it has the consistency of honey; then put it into a mold and leave it for 3 days. Then take it out and you will have a seal of gold better than the real thing. To avoid being called dishonest, keep the recipe secret.

53. The liquefying of gold for a painting

Orpiment and cuttle-fish bones and the efflorescence of copper, in equal portions: sandarac, gold-colored litharge, and egg yolks, in equal portions: with the above grind gum-tragacanth and goat gall. When you have mixed this with the gall liquid alone, use it on a strip on which you impress a seal. Run through this the materials that are to be coated. Make your seals with a filed iron [die] that is burnished, not rough, and coat it with smoke from burning incense. It will be a marvel.

54. Making [figurines of] green [and red] gold [and silver]

Take 4 parts of gold, 2 parts of silver, and melt them together; and when you have melted them, make it into one or another male figure that you want, and you will have a manly color—no mean display, and a delight that affords to living men the color of living figures.

---

[28] In *L* the units, in this chapter but nowhere else, are denoted by γ supposedly denarii as in *P*, chapters 276 and 277. This chapter is not in *S*, It is a highly corrupt version of No. 74 in Leyden X.

But if you want to make it red, mix 1 part of Cyprian copper. Melt prime copper a number of times until it becomes brick-colored, and melt it with the above-mentioned weight [of gold].

Now, if you want to mold a female figurine, take 1 part [of gold] and 4 weights of silver, and the mixture is made, showing a female body that gleams when it is polished. After this it has been found how there may also be made black figurines of gods from gold, silver, copper, and other alloys. The mixture and the operations will be shown in the following.

55. The gilding of everything that you want to gild, whether it be a vessel of silver or copper

Take an extremely thin gold sheet, shear it into tiny pieces and put it in a mortar. Add a little quicksilver and leave it for a short time. Afterwards, add some natron and vinegar; rub it thoroughly with a pumice stone until it has the consistency of glue on account of the abundance of quicksilver. And now you put it in a clean cloth and squeeze it, so that most of the quicksilver comes out. Then you take the vessel [that is to be gilded], polish it with fine pumice, heat it, and while it cools coat it with the amalgam, and you heat the vessel a second time and again coat it and put it on the fire. And the gold alone becomes enriched. After a little when the color pleases you, heat the vessel and put it into a blacking liquid, i.e., the boot dressing, with which leather things are blackened. And then rub it. But if you were gilding a copper vessel, after you have polished it, coat it with liquid alum, for it will [then] receive the amalgam.

56. Painting black on a gold vessel, so that you think it is inlaid

Melt together equal parts of silver, red copper, and lead, and sprinkle native sulphur over it. When you have cast it, leave it to cool, put it in a mortar, grind it, add vinegar and make it the consistency of the ink with which writing is done. Write whatever you wish on gold and silver [vessels]. And when it has hardened, heat it and it will be [as if] inlaid. Melt it like this: carve charcoal and so put the silver and the copper in it and melt them (and while you are heating them add in the lead, then the sulphur).[29] When you have mixed it, pour it out and do as was said above.[30]

57. [Decorated work]

Take the stone androdamas,[31] grind it and mix in some chrysocolla and apply both [to the vessel]. Put it in the furnace and wait till it is well cooked, and it will be decorated.

58. [Niello]

Melt silver and when it is red hot add sulphur, and stir. Then leave it to cool. Grind it well; coat [the vessel] with it, adding natron and oil. All these things are uno veda[32] before they become anything.

59. To show ornamented gold and ornamented work of copper

Take 3 parts of the best gold, 4 parts of Macedonian chrysocolla, 1 part of efflorescence of copper, 3 parts of the best silver, and add the preparation. Sprinkle it with a little reddish natron, and melt it over a slow fire. And, when you have mixed it into the ground chrysocolla, take it out and work vigorously, making whatever kind of work and stamped design you want.

60. The gilding of tin sheets

Take leaves of tin, dip them in vinegar and alum, and glue them together with glue made from parchment. Then take saffron and pure (i.e., clear and transparent) glue, drench them both in water with vinegar and cook them with filings [parchment scrapings?] over a slow fire. When the glue shines, coat the tin leaves [with it] and they will appear golden to you. But be careful not to add elidrium. Now, if you have everything already ground, do not add glue, for your work will be hardened. If it has been stiffened with glue, for gold writing add elenusia, so that you may extend it.[33]

61. An easy way of gilding

Grind gold-colored sandarac and fissile orpiment and gum-tragacanth with goat gall and the inner parts of an egg. First coat the new work with oil. When it is dry, it will have a color like gold at its best.

---

[29] The words in parenthesis are interpolated in a later hand.

[30] This recipe, which is repeated almost verbatim in chap. 206, describes the making of niello, but it is applied as convex writing in relief on the gold or silver surfaces. Such treatment is rare in Europe, where niello was more commonly melted into engraved lines or champlevé areas and polished level to contrast with the ground in color only. Sasanian and Byzantine silver with niello in fairly high relief is known, but the metal was prepared by coarse chisel cuts to help anchor the niello. This is quite different from the "ink" described here.

[31] Pliny (36 Para. 146) cites one Sotacus to the effect that androdamas is a black hematite. If it were marcasite, this recipe might give an iron-copper sulphide as an inferior niello. Chrysocolla is presumably gold solder, i.e., verdigris or some other copper compound for use as in chapter 117, but if it were borax or if some other glass-former were present the mixture would make a fair glaze for decorating pottery.

[32] Meaning unknown.

[33] Chapter 204 is a duplicate of the first half of this chapter, except for reading effluxerit for effulserit. Elenusia may perhaps be linseed oil.

62. *Coating surfaces*[34]

Burn gum and grind it with water and coat the material. Then place them on a smooth stone and put it into the fire. Smear them with the above-described preparation, and their surfaces will be coated.

63. *Again*

Coat the pieces lavishly with ceruse and Greek natron together with ox glue.

64. *The application of gold to iron*

Heat the iron gently and smear resin on until it smokes, and place [the gold] on. When you have wrapped a cloth around the pure stuff [gold] on the iron, bind it.

65. *For gold to increase*

7 oz. of copper, 6 oz. of gold-colored orpiment.

66. *Making gold*

6 oz. of copper.

67. *The duplication of gold*

1 part of silver filings. Italian chrysography from a woolen polishing cloth. The gilding will be easy, if you do the same things.

(A liquefaction of gold: put ox gall in a new pot, and leave it in the new pot for 6 days; next, wet it and, when you take it out, join it and mix.)[35]

68. *[Writing in gold]*

Take gold and put it in a new pot, and add ox gall. Grind it for 3 days and, if you see it has become liquid, gradually pour out the gall, and add fresh gall with vinegar. Then transfer the pure [gold] again into a copper pot; and when you have warmed the pot, wash off the pure [gold] and dry it. Coat the surface with water-glue, and so write. But if it is not made liquid within three days, leave it [to grind] longer, and it will become liquid.

69. *[Giving to glass the nature of a stronger metal]*[36]

Take a sound glass pot, hollow, in the shape of a mortar, and scraping it with an emery stone, scratch all its curved surfaces in a crisscross fashion. Put into a melting furnace some ashes from which glass is to be melted, and sprinkle them with raw dragonsblood. Now, if you do not have raw dragonsblood, make a preparation from the white of eggs and mistletoe juice, and sprinkle the ash in this blood, and then cook it. Now, after it is cooked, when you want to produce in glass the shape into which it is to be formed, coat the vessel with the same blood. When that is so done, you should know that fragile glass is formed into the nature of a stronger metal. Yet you can break it in this way. Take the blood of a cock and grind with it celidon-stone, mixing in some urine, but not using all the urine of one man. Then add gum tragacanth, and when they are all mixed together, put them into that glass pot. You will see that *Arte et Ingenio vinci Ingenium*.[37] Again, if you put this itself [the glass] into a lead or tin vessel, it becomes solid like copper or iron, even though you don't use cock's blood.

70. *The same [gold] used as ornamentation*

Make a pot rough, as if it had been filed; and put Attic honey in it and with it coat gold that you have hammered out into the shape of a straw. Lightly pass your hand round it, and afterwards grind it until it has become smooth. Then after all the gold, or as much of it as you want, has been ground, add water to the honey and wash it delicately; when the honey has become liquid, you will find that the liquid gold settles; then use it as a coat. So, add sufficient ox glue, or fish glue, or gum, to that which has settled, which you should have in a pot for use. With this write letters and make seals. Coat and paint roofs with this, and everything that you want will look like gold, whether you paint or you write. When it is

---

[34] The word translated as coating, *conjunctio*, usually means the joining or fitting together of two separate objects. Here, as also in chapter 64, it clearly refers to the application of a protective surface coating or varnish.

[35] This paragraph is interpolated at the foot of the page in a later hand. After this point in the manuscript the scribe ceased to apply numbers to the chapters and all enumeration is that supplied by Phillipps or his editor Way in 1847. (See note in introduction, page 9.)

[36] This chapter, which is untitled and obviously out of place sandwiched between two chapters on chrysography, is exceed-

ingly obscure, but it is possible that it refers to the devitrification and hardening of glass that occurs when it is subjected to a long annealing process at a moderately high temperature. This was first studied systematically by R. A. F. de Réaumur in his paper, "L'Art de faire une nouvelle sorte de Porcelaine, . . . ou de transformer le verre en porcelaine," *Mem. Acad. Sci.* 1739 (published 1741): pp. 370–388, but it must have been repeatedly observed by artisans who used glass vessels in distillation, sublimation or other processes involving prolonged heating at intermediate temperatures. The outer layer of lute composed of ash and blood would prevent the hot glass from sagging before it had hardened, and also serve to nucleate the crystallization on the surface, which would be aided by the scratching. Note also the suggested use of lead or tin, supposedly as a molten bath which would both heat and support the glass. For some subsequent history of the process and its influence on geological thinking, see C.S. Smith (1969).

One of the chapter titles listed on fol. 5 of the Sélestat manuscript reads *Vitrum quod non frangitur facere*, but it is not present in the text and none of its associated chapter titles resemble any of the neighbors of the present one. Our title comes from the text itself.

[37] An epigram that can be loosely translated "Art and imagination subjugate the qualities of matter."

dry, rub it with a tooth, so that the coating looks brilliant. By this plan, indeed, you also liquefy iron, lead, and copper, and write and coat whatever you want.

71. [Writing in gold]

Melt lead a number of times and pour it into water; and, when you have done this many times, remove the lead. In the same water into which you have been pouring the lead, [pour] melted gold, and leave it there. For it breaks into tiny pieces; take these, grind them and they liquefy. To this add as much of the above-mentioned glue as you want; and use it for whatever you want. Now, indeed, the gum that is found in the deepest hollows of a tree holds the gold well if it is previously smeared onto glass vessels that are being gilded.[38]

72. [A recipe for silver]

Take Cyprian misy, sandarac and *elidrium* in equal portions; and with water in which should be cooked the leaves of mountain sandarac, i.e., of wild poppy,[39] which is also called *ammoniacum*; make a [paste] the consistency of glue, and melt silver, but the best silver. When it is hot, dip it into the above-mentioned water.

73. A recipe for silver

2 parts of Cyprian copper, 1 part of silver, 4 drams of sal ammoniac and the same amount of fissile alum and licorice. Melt them all together. Now if you want to work with this, you take the stuff squeezed out of thistle and ground raisins, and you put little lumps of this in a pot and cook very well, take them out and work by the fire. Dip [the silver] in the preparation that comes from the cooking.

74. [Making brass]

Take ductile copper of the kind which is called *caldarium*,[40] or fired copper, and hammer it out and make sheets of it. Spread these sheets out on the bottom [of a pot] and sprinkle on them thoroughly ground white cadmia. This is engendered in Dalmatia, where coppersmiths use it. Lute the furnace thoroughly with potter's clay so that it cannot breathe for a day.[41] Then open up and if it is good, use it; if not, cook it with cadmia a second time, as above. If it comes out well, the *caldarium* copper is permeated throughout with gold [color].

75. Making copper white

When [copper] begins to melt, add orpiment, not prepared, but fresh.

76. For the dyeing of gold

When it begins to melt, put in orpiment that has been prepared.

77. Making copper white

When copper has been melted, add unprepared orpiment and it becomes white, and something which can be ground [to powder]. Wash this fairly often with water, until it becomes clean; and take some of it, and you will find it to be yellow. Then wash it off with water and you will find copper like blood. To this add silver in the furnace and it turns into a silver like coral. Mix 1 part of this and 2 parts of gold, and you make a marvel.

78. Making lead like silver

Put into a cooking-pot 1 *mina* of lead, some efflorescence of copper and 4 drams of ground pomegranate skin. Coat it with chalk and leave it until it is melted.

79. A recipe for gold solder

Melt together 1 dram of lead and 4 drams of copper, adding Samian earth, salt, and liquid alum and let them liquefy; and as soon as it begins to do so, clean it with a [flux] preparation. Now, if it can be extended in the fire, i.e., if it is *caldarium*,[42] dip it in vinegar and cast it into whatever shape you want. It will become extremely white.

80. Changing copper

Take 6 *minas* of salt and 4 *minas* of filings or scrapings [of copper]. Mix the filings in a pot with ground salt, sprinkling vinegar over them, and leave for 3 days and you will find it has turned green.

---

[38] *Cf.* chapter 21.

[39] This phrase, on the first line of fol. 9 *verso*, is repeated in a marginal gloss in a later hand at the bottom of 9 *recto*, with another note defining sandarac as varnish gum.

[40] Note that *caldarium* copper (literally copper for kettles) is here described as fired, i.e., refined, copper in ductile sheets. This would be proper for making brass, and in his detailed account of making brass Theophilus (Book III chapters 64 to 67) specifically calls for refined copper as the basis of the best brass for gilding, though not for the coarser kind used for casting into cauldrons and kettles. Pliny, however, says that *caldarium* copper is brittle and suitable only for making castings. To further confuse matters, in *Mappa* chapter 79 the ability to be extended in the fire is used as a test for *caldarium* copper (showing that it is not hot-short), but it is made by alloying copper with some earths and 20 per cent lead which would surely embrittle it, especially when hot!

[41] An earlier version from which this is faultily derived more plausibly says to melt it for a day, taking care that it does not evaporate (Berthelot, 1887, transl.: p. 292).

[42] This chapter seems to be unrelated to its title. On *caldarium* copper, see note to chapter 74.

81. [Writing in golden-colored letters][43]

Grind 4 drams of litharge with dove feces and vinegar. After heating it, write with a stylus.

82. Writing in silvery letters

Grind quicksilver and lathe-turnings of tin; pour in some liquid vinegar and, when you have ground it, make it the proper consistency for writing.

83. Making silver, elidrium, or gold from copper

Take 4 parts of copper and 1 part of silver, melt them together and add 4 parts of unburnt orpiment, i.e., eureos and 1 part of silver. When you have thoroughly heated them, let the mixture cool and put it in a pan. Coat it with potter's clay, then roast it until it becomes cherry-red and melt it, and you will find silver.

But if you roast it a lot, it will become electrum.[44] And if you add to it one part of gold, it becomes the best gold.

84. How black silver should be made white

Take 2 solidi of minium [i.e., cinnabar] from which quicksilver is made, which is washed in hot water and, when dried, is ready for use. Melt it, and mix in a little bit of copper with silver and lead. Now, all of them accept lead [when] they are melted, and so a more resplendent sign is achieved.

85. The alloy of silver

Grind calf gall, a woman's urine, the seed of rue, a third part, and likewise a third part of male madder; then pour in a third part, and pour it into silver, and place it on the top of the furnace. Spread it about until it becomes hot, and then put it in the inside of the furnace.

85-A. [A copper-tin alloy ingot]

1 part of Cyprian copper and 1 part of tin should be melted together into an antabra. An antabra [an ingot?] is a shape used in a mint.

85-B. [A white silver alloy]

2 parts of silver, 3 parts of purified tin. (Tin is purified like this: it is melted with an admixture of pitch and bitumen.) And melt together with them half a part of white copper. Then take it out and grind it and make whatever you want.

86. Cleaning silver vessels without damage

Take a dirty woolen cloth, soaked in thick salt water, and wipe [the vessels] clean; wash them off in cold water and let them dry. Wiping them clean with this causes no damage.

86-A. [Again, cleaning silver]

With sap grind red natron, not entirely pure alum, and sea-leek sprouts with the juice of a lupine cluster. Coat the silver with it using a stiff feather; heat it in a pot in which there has also been quicksilver.

86-B. [Again, cleaning silver]

Melt red natron and alum together. Then grind it with water, coat the silver, and heat it.

86-C. [A recipe]

Take half a mina of Cyprian copper, and 1 mina of tin, 8 oz. of magnesia, 20 oz. of ground pyrite, melt them together, and put in the tin; then, not last, add some quicksilver, and stir with an iron rod. Cast it into little cylinders.

86-D. [To clean silver]

Grind minium with alum, pour in white vinegar, and make it the consistency of a wax salve. And when you have coated [a silver vessel with it] several times, leave it in that condition the whole night through.

87. For silver to appear golden colored

Minium, liquid alum, Cimolian earth; equal amounts. Pour these into sea water, and after thoroughly heating dip the silver in it.

88. The recipe for white [metal]

Take 4 drams of white [copper] filings, and quicksilver; and without adding anything, sprinkle them with granulated alum, and melt.

89. The recipe for liquid silver, by means of which one plates copper[45] with silver

Liquefy in salt 4 drams of silver filings, 15 drams of Attic honey, 5 drams of liquid resin, 2 drams of burnt copper. Then put it in a box.

89-A. [A recipe]

1 pound of cleaned tin, 1 dram of quicksilver, 1 dram of Brindisian mirror metal,[46] pounded and sifted.

---

[43] Heading erroneously reads Item.
[44] Probably a scribe's error for elidrium. But elidrium may sometimes have been a kind of electrum. See also chaps. 15 and 209 for almost identical recipes.

[45] Text reads aurum but we find ourselves unable to follow it!
[46] Supposedly this a true brittle speculum metal containing about 33 per cent tin, though this may refer to the Brindisian leaded bronzes of chapters 221-A and B.

The quicksilver should be mixed with the tin, but the others should be melted separately and [then] mixed. And when it has become liquid, thoroughly heat copper, and when it has settled and you have smeared it, dip [the vessel] in that liquid and, just as waters become stagnant, heat the mouth of the top of the little vessel, and it will not come off.

89-B.   [Black inscriptions on silver]

Grind a little burnt lead in a mortar, mixing in some sulphur and with vinegar make it the consistency of glue.   Inscribe silver vessels [with this] and, when it has dried, heat them and it will never be worn off.[47]

89-C.   [A tin-copper-lead alloy]

Grind all together 1 part of copper, 5 parts of tin, 1 part of lead.

89-D.   [Treatment for copper vessels]

Grind together into one, chrysocolla, quicksilver, reddish Samian earth, and honey.   Coat a copper vessel with it and roast it.

90.   Silver of a blue color

Now, there is a potion compounded like this: you cook down fissile alum in water, add quicksilver, and 1 part of roasted basanite stone,[48] 1 part of sheet silver.   These are made so that you can fashion whatever potion you want, and leave it to dry for 3 days. Nor should you fashion it first, not to tell you what it is, for it will be destroyed, since it will become like sand.   Now heat it, applying it only once; for if you want to do it twice, you lose the mixture.   Fashion seals of it, not unlike a greenish blue.   Now if you want the same things to be gilt [i.e., golden colored], mix *chrysitis* and saffron with liquid gum.

91.   How silver and copper may take on the color of gold

1 part of gold, 1 part of lead: melt these together, then file them and grind in a hard Theban mortar, adding water and natron.   Afterwards pour them into a lead pot.   Then you dip the vessel [to be gilded] in the mixture, put it into the furnace and heat it until it has the color of gold.[49]

91-A.   [A metallic ink]

Take 2 drams of *stagnum*,[50] and 2 drams of tin and

melt them.   When they are thoroughly mixed, each will lose its character and a fragile material results. You grind this in a hard mortar, adding gum, and wash it off; dry in any way.   When it has dried, smear it on, as suggested.   Now, if you want it to be of a golden color, mix ground saffron with clean glue; write what you want to delineate; when you have written and it has dried, rub it with a burnisher.

92.   Silver writing in the Italian way

Take silver sheet; grind it, as described below, with salt or natron.   Then wash it off with water, add ox gall and, grinding it together in a glass pot, lay it aside.   Write with a quill or with a paint brush; when it has dried, polish it.

92-A.   Silver writing in the Italian way

Take silver sheets, grind them with quicksilver, *stilbada*, alum, gum, and vinegar, and so write.   In order that the writing may last, mix blacksmith's water[51] with every pigment and the same amount of alum, and so write, after you have previously carefully wiped clean [the surface to be written on].

92-B.   [A green writing]

Put into vinegar, efflorescence of copper, verdigris, and sulphur.   Add gum and write.

92-C.   [A silver writing]

Grind silver in a mortar [made of] basanite stone with a little water, and liquefy it with [more] water, and it will dry out as ground silver.   Taking on water-glue, write with it.   And if you are working in gold, use it in the same way.

92-D.   [Another silver writing]

Melt 1 dram of silver, and when it is melted, add 3 drams of pure tin, and pour it out and leave it to cool.   Then file and grind it, and write whatever you want.

92-E.   [A leaden writing]

Melt together 2 parts of tin and 3 parts of lead. When you have done this, polish it, file it, and grind it.   Then add water-glue, and polish.

93.   Writing with quicksilver

Take quicksilver and put it in a porringer; add a little quicklime and a little liquid alum, and some very sharp vinegar.   Stir it until it becomes quite uniform.   Take it off the hearth, grind it and put it in a cloth.   Squeeze it and quicksilver will drop down. Add some isinglass and write.

---

[47] Cf. the superficial niello of chapter 56.

[48] A hard fine-grained abrasive stone used for whetstones, mortars, and the touchstones of the gold assayer (Pliny).

[49] An interesting variant of gilding with mercury. It would be very hard to oxidize the lead away without diffusing silver from beneath.

[50] *Stagnum*, clearly here an alloy of tin, also means the pure metal. Elsewhere in the manuscript tin is *stannum*.

[51] Water in which red-hot iron has been quenched (Pliny).

93-A. [A copper composition]

2 drams of Cyprian copper and 1 dram of white magnesia, 4 drams of litharge.

93-B. [A substitute for silver]

Take some thin white tin, purge it [by drossing] 4 times, and melt it with 1 part of silver. When you have melted it, grind it thoroughly, and fashion whatever you want, cups or whatever seems good to you, for, even to master smiths, it will look like prime silver.

93-C. [Metallic writing]

Take pure tin, as you have learned: 1 dram of tin, 3 scruples of magnesia; clean off the soft dross. File 3 scruples of it and 3 scruples of quicksilver. Afterwards take iron filings, grind them in a mortar, until they become wooly. Then add an amount of magnesia (and grind these carefully until they become amalgamated). Put this in the furnace, and melt it; then you put this, together with tin, into another furnace. Learn this [way of] writing.

94. Using copper to make it like silver

Grind 2 drams each of chrysocolla, earthy ceruse and quicksilver, and pour in a sufficient amount of good honey, and heat it. While you are coating on what you want, after you have first wiped the vessel clean, lightly hold it over a fire of white poplar wood, i.e., the violet kind.

Smear on quicksilver and sulphur; and there will be 1 *mina* of Cyprian copper vitriol, 1 pound of roasted misy, 14 drams of common alum. Take in your hand fine salt and hold it in your hand until it becomes black. Thoroughly wipe and rub a stater with salt. And when it has become silvery, from a copper stater, wrap in vine leaves a pennyweight *duci*; and leave it in this condition for the whole night. On the following day, take it out and use it.[52]

95. Making green letters that last for ever on copper, wood, stone or whatever you wish[53]

Take filings of *naxia* shaved from the skin and with them grind *agacia* filings, and alum, and water from a lake, i.e., rainwater, and the leaves of barley that has gone to seed. Grind them all together, write, and the letters will be green.

95-A. [A coating for copper]

Wipe some copper and rub it thoroughly with pumice stone and leave it in the sun. Afterwards,

take 1 part of citrus wood[54] and pieces of pomegranate wood, cut them up carefully and take a convenient amount of them mixing them with natron, water, and salt. Put in the copper and leave it there for 5 days, and when it is dyed, coat it with wax salve.

95-B. [A coating for copper]

Grind 1 part of the sap of a *columbaris* thorn with alum; pound it thoroughly, gather it up and leave it for 2 days. Then when you have taken it up, coat [copper?] with it, leave it for 1 day and then wipe it off.

95-C. [To solder copper with silver solder]

Take 1 dram of silver, 1 dram of copper, half [a dram] of lead and melt them together. Now, when you want to do the soldering, apply the solder to the copper, and, when the [soldering] iron is hot, put it on to the solder that lies on the copper, and so decorate. Otherwise, join a sheet to a sheet.[55]

96. Making green pigment

Coat copper, beaten out into sheets, with honey or the froth of cooked honey and put beneath it in a pot broad laths of wood, and pour over it a man's urine. Let it stand, covered, for 14 days.

97. Making indigo pigment

Collect the juice of dwarf elderberries and dry it thoroughly in the sun. From what remains make pastilles with a little vinegar and wine, then use it.

98. Making Greek glue

Make a flour out of varnish by grinding it on marble and sifting it. Put it into a rough cooking pot, thoroughly sealed with a cover, so made that there is a small hole in the middle of the cover and in this hole a pointed iron rod. Now, put it on iron sheets on a goldsmith's forge, the fire in which should have been previously started. Then put dry sticks of wood, cut up very small underneath [the pot] and [the varnish flour] melts as soon as it grows hot. Take out the iron rod and put a little drop on your fingernail; and if it seems liquid, take it off the fire and pour on 2 parts of oil squeezed from flaxseed [i.e. linseed oil] to 1 part of varnish, and again put wood underneath and cook a short hour, and use it. But if there is a grain of mastic [in it], it liquefies more slowly.

---

[52] Latin unclear: a fraudulent operation seems to be intended, in which a stater made of copper is superficially silvered.

[53] In the manuscript this chapter title appears in the text without the usual distinction.

[54] Not our citrus but a fragrant African wood highly prized in the ancient world.

[55] This paragraph describes the use of a hot soldering iron to melt silver solder onto the surface of copper, apparently in a decorative design. Biringuccio (1540) also describes the use of glowing-hot tongs to melt silver solder (in joining a broken sickle, saw, or sword), but an iron would not today be used with a solder of such a high melting point.

99. Gilding on stone, wood, or glass

Now, the man who is gilding glass should take a piece of pitch glue and a piece of almond gum, mix them, cook them and coat the [glass] vessel; and cut up gold leaf very finely and lay it on to form the picture that you want to make. And the same for stone, after washing it in water, and for wood. And, when the glueing has dried, burnish it with a hematite stone or with a [burnishing] tool.

100. For coloring gold

Take vitriol and roast it, as you know how, and as much salt again, and temper it with red wine, not too thinly, in a copper pot, and coat the gold with it. Put it in the furnace and heat it until it becomes black, then take it out.

101. The layout of a structure

The layout of a structure for those who set down [S] either with what measurements you should lay out [the plan of] buildings or with what measurements you should raise them in height, depending on the method of construction.

If it is to be 4 times the height of a man, the foundation should be constructed to one man's height. But if the height is to be 3 times the height of a man, the foundation will reach as far as the crotch. And if the height is that of one man, the foundation will reach as far as the knee. If it is tile work [add] four cubits [S]. If it is roofed in wood, it is included in the height. If it is a vaulted roof, you must excavate the foundation as deeply as the height; namely, the height measured ought to be the same as the wall without the arch.

Now if the place is hard and of solid rock, lay the foundation one cubit less for each man's height. If the place is soft, build as we said above. If the place is stony, do not trust the stones, but excavate as one should [in soft ground], so that it is not pressed down by the excessive weight allowing the structure to subside.

102. A structure in water

If it is necessary to erect a structure in water, make a triangular coffer and seal it outside with tallow and pitch, so that the water may not enter and wash away the mortar and the men who are working inside. Put the coffer between four ships, and fix it firmly in the place where it should be; the ships themselves must be anchored[56] so that they do not move in the

water. Then lay the stones for building the structure. Now, the tempering of the lime should be as follows: Put together 1 part of sand and 2 parts of lime and then work it. The coffer itself should be one cubit higher than the water.

103. Mortar

This is the way mortar should be made. Put 1 part of lime, 3 or 4 parts of sand, a third [of a part] of crushed tile, a sixth part of pulverized chaff, also one congius[57] of water and 2 sextaria of hog fat. Let it rest for a week—it will improve if you leave it longer. Now wet it continually according to the amount that it needs and it will be compounded. Then work with it.

104. Licamonia

Licamonia: Three solidi of bits of Egyptian alum, 1 oz. of natron.

105. The recipe for cinnabar

Take 2 parts of clean quicksilver[58] and 1 part of native sulphur, and put them in a flask, and, cooking them without smoke and over a slow fire, make cinnabar. Wash it properly.

106. The recipe for verdigris

Take very clean copper leaf and hang it over very sharp vinegar. Leave it undisturbed in the sun for 14 days. Open it up, take away the leaf and collect the efflorescence; and you will make the cleanest verdigris.

107. The recipe for white lead

Take lead, make a leaf and hang it over vinegar. Collect the efflorescence and wash it well until it is clean and you will make white lead.

107-A. The recipe for a Pandius

Afterwards, take 1 part of cinnabar, half a part of verdigris and half a part of white lead; put them into a marble mortar and grind them well. Now, after grinding, take some water in which fish-glue[59] is cooked, and it will become a pandius pigment.[60]

---

[56] Following S, hornizas. P reads oneratas, L, horizas. Chapters 102 and 3 are present in all three manuscripts, L, S & P, but they were omitted in Muratori's transcript of L as they are on a leaf (211v) somewhat separate from the main body of the chemical recipes. Svennung, who believed them to be his discovery, published them in 1941, but Burnam (1920) had previously done so.

[57] The congius was a liquid measure of about 3.6 quarts; the sextarius, one sixth of this.

[58] Ydroargirum, literally water silver. Elsewhere quicksilver is argentum vivum. Pliny uses the word hydrargyrum for artificially prepared quicksilver. See chapter 221-C for a more detailed cinnabar recipe.

[59] Theophilus (I.28) gives a good account of the making of fish glue, ichthyocollon (here icciocollon). For its use in painting see Roosen-Runge, 1967: 2: p. 7 and passim.

[60] In chapter x, a similar mixed flesh-color pigment is called carnatura. It is what Theophilus calls membrana. The word pandius appears numerous times in the manuscript in different contexts (see index). It is usually a pigment and, as in chapters

108. *The recipe for azure*

Gather leaves of the violet flower and grind them well in a clean mortar; add soap made from axle grease without using lime. Make it clean, scale it off with warm water, and dissolve 1 oz. of soap to a pound of water, and rub the soap very finely with the water and leave it to cool. Afterwards put this mixture in the ground flowers and put it in a glass pot which can hold it and lay it aside. And after some time mix and stir it every day once a day until the end of the week. After this, leave it for 3 days then stir it for 2, until it is cooked down. Next, take the greater dark lily which is purple and has little knife-like leaves. Rub it down similarly in a mortar as usual and leave it without soap, adding water. Next, put [together] 2 pounds of the violet composition, 1 pound of the greater dark lily, and some frothed Egyptian alum, [or] if there does not happen to be some frothed alum or, if it is weak and raw, put in 2 ounces of soap; and 2 pounds of frothed urine. Cook it all down over a slow fire for 6 hours, and if it is too green, add urine; if too blue, add more alum.

Now if the azure is viscid, add a sufficient amount from the white domestic lily, and cook it down.[61] Now inspect a sample of the concoction on the stick with which it is stirred, for it does not show the color when it is hot, but when cold it does. And the cooking ought to be over a slow fire. For, while it is being cooked, it will lose its water; therefore more water should be mixed with soap, according to the proportion we describe above, and added to the concoction. Now when you take the glazed[62] earthenware pot out

of the furnace, the water tends to rise to the top of the mixture.

Now alum should be freed from froth in strong soap. Put the alum in warm water, and let the alum settle, pour off the warm water, and so de-froth the alum.[63]

Then grind this [azure] mixture down so that it is reduced to powder and there are no little stones in the azure. For after the cooking, the mixture must be ground and stirred and dried in the shade in the same pot. After that, it should be thrown out into the sun and baked so that it becomes azure.

109. *How [to treat] a painting so that it cannot be destroyed by water*[64]

Coat a painting in the sun with the oil called castor oil, and it is fixed so tightly that it can never be destroyed.

110. *The recipe for a* pandius

Gather the leaves of the flowers of the black poppy and put them in a new quill. Cover and put in the sun for a day. When they are dried out, take the water in which fish glue is cooked, and put the leaves of the flowers in it and grind it properly, mixing it with a little cinnabar, and the *pandius* color will emerge.

111. *The recipe for* ficarin

Take clear lac and grind it cleanly and cook it in defrothed urine and pour the broth that emerges into a glass pot. Afterwards take well-ground, dry powder of the white domestic lily. Now if it begins to darken, do not put more in for fear that it will become too dark. When this is ground, mix it into 2 pounds of iris flower, and when these are ground mix the two together into 3 pounds of [lily] powder. Then for each cooking of the lac mix in 1 oz. of Egyptian alum and grind well. Put them in a new glazed earthenware pot and set them out to warm a little, but not to burn; afterwards mix the concoctions into the lac and make them boil. Take them from the furnace and dry in the sun.[65]

112. *Gilding on wood or on cloth*

If gilding is to be done on wood, steep almond gum for one day. Next grind the gum properly with water, add sufficient saffron and dip [the wood] into the water with the gum, warm everything over a slow

---

107-A, 150, 176 to 189, commonly contains cinnabar and white lead. Sometimes alum and natron are incorporated. In 192-D the statement occurs that "Ochery earth is a *pandius*—you can color everything with it, compound everything with it." But in chapters 235 to 239 it is also used for the dyeing of skins. When detailed recipes are given it is always a compound pigment, though it occurs in many different colors (green, purple, flesh-color, and like ocher or cinnabar). In chapter 215 it seems to describe the color of a natural variegated rock or mineral efflorescence. Roosen-Runge, 1967, 2: p. 66, suggests that the word pandius is a latinization of the Greek adjective *pantoios*, "of every kind," and that it means a *mannifältige Riehe* or complex series of pigments. It evidently is used in the *Mappae Clavicula* to refer to a combination of compatible pigments or dyes, the proportions of which can be continuously varied to produce a gamut of intermediate shades of color as desired. We know of no English word that properly conveys the idea.

[61] In *S*, the balance of this chapter was first omitted by the copyist, but a separate small piece of vellum has been inserted which contains it and chapter 109 written in another, apparently contemporary hand. This is bound in sideways, though numbered consecutively folio 16. The verso is blank.

[62] *Bituminatus*, literally bitumen-coated, to make it watertight. However, ordinary ceramic glazing is clearly meant, for the potter's recipe for *bituminatio* in chapter 143 specifies the use of lead oxide, and bitumen would be a poor seal against some of the oily incendiary compounds for which *bituminatus* pots are used in chapter 275 and elsewhere. A white marzacotta glaze based on lead silicate with tin oxide appears in 281.

[63] This process, variously called frothing or defrothing—*spumato* or *exspumato*—is evidently a clarification procedure, used on urine, alum, and other materials in lieu of filtration.

[64] In *S* this bears the heading *Ad pingendam picturam Hudis*, "to paint a picture against dampness." For a better varnish see chapter 247.

[65] Another recipe for *ficarin* occurs in chapter 223. Lac was sometimes kermes. See also Roosen-Runge, 1967: 2: p. 32.

fire and work on the wood whenever necessary. On cloths, however, or on walls, take thin egg-white, add sufficient saffron, dip [the cloth] in and when [all] is mixed and ground, put it aside in a glass pot.

Again, mix 1 oz. of linseed oil, 1 oz. of steeped gum and sufficient saffron: cook down with water.[66]

The following three chapters [are for use] when it is necessary to work in gilding with leaf.

113. The recipe for linseed oil

2 pounds of linseed oil, 1 ounce of gum, 1 ounce of pine resin. Grind all these and cook down in an earthenware pot.

113-A. Linseed oil for gilding [S]

2 pounds of linseed oil, 2 ounces of gum, 1 ounce of resin, 2 solidi of saffron. Mix these three as above.

114. The procedure for laying out [gilding] [S]

If gold leaf is to be laid out on a firmly stretched [S] raw skin coated with white lead or any pigment, the gold leaves are laid down and after they are dry, coat them with linseed oil using the mixture described above, where we say it is mixed with saffron.

115. The application of leaf gilding

Leaves should be made of tin. They should be made like this: melt the tin well and slowly pour it out onto a marble slab and make thin leaves of it, just as if they were of gold.[67] And apply them in the same way as gold leaves, as we taught above. Cook down the herb swallowwort and add to 3 ounces of this concoction, when it is strained, 3 solidi of saffron and 1 solidus of orpiment. [Coat the tin leaves with this.]

116. Coloring tin leaf

Take 1 ounce of clean saffron, 2 oz. of the best split orpiment [S], add half [an ounce] of gum and a

half ounce of linseed oil. Mix in rain or fresh water, and boil it. Mix the preparation together, grinding it well, and taking it up with a sponge, coat the leaf with it. When it has dried, coat it a second time, and when it is dry, rub it with an onyx until it shines brilliantly.

117. The recipe for gold solder

1 ounce of copper calcine,[68] 3 solidi of olive-oil soap, 1 solidus of calcothar. For use, mix these together, first grinding the copper calcine and the calcothar separately into powder. Mix with as much soap and water as is necessary for gold solder.

118. Another gold solder

1 pound of copper calcine, 2 solidi of alum.[69]

119. Again

Gold mixed with quicksilver is put into a furnace until the quicksilver itself burns. Afterwards, take out the gold and grind it in a mortar, until it becomes powder. Mix this with olive-oil soap, as much as is sufficient for the composition of gold solder.[70]

120. Silver solder [S]

2 parts of silver, 1 part of copper.

121. Again, another silver solder [S][71]

Put silver mixed with quicksilver on the fire until the quicksilver itself becomes dry. Then for use, grind it until it becomes powder; mix it with soap and water, as much as is enough.

---

[66] In view of the later importance of linseed oil to the artist it is interesting to note its first appearance in this and the following chapters which are present in all three manuscripts: *L, S* & *P*. Although linseed oil is here used as an adhesive and base for gold leaf rather than as a vehicle for pigments, its virtues for the latter purpose would, of course, be easy to discover once its use with metals had brought it into the painter's environment. Indeed, the orpiment-loaded color for tin leaf in chapter 116 is a true paint, though it is used to produce a uniform coating rather than for brush-applied detail. See also chapters 98 and 246, both of which refer to the drying action of an admixture of mastic.

[67] It is hard to see how a thinness in any way comparable to gold leaf could be obtained by casting. Possibly, the marble slab was sloped and the molten tin, fairly hot, run over it as was later done in casting sheets of tin and lead for organ pipes and roofing. More probably, a hammering operation followed the casting. This is mentioned in chapter 207, which is a slightly different version of the same recipe.

[68] *Caucucecaumenum* [P], *calcucecaumenum* [S]. This is a Latin transliteration of the Greek words *chalkos kekaumenos* (roasted copper) which is more commonly *aes ustum* in Latin. It was copper oxide. Chapter 139 tells how to make it by roasting the sulphide. Its use as a solder depends on its reduction to metallic copper by the soap or in a reducing flame and superficial alloying with the gold to give a liquid alloy. On fine work this method of soldering is to be preferred to the use of a pre-melted alloy, for it can be painted on in small amounts and better controlled. It was probably the basis of the superb granulation work of the Etruscan goldsmiths. Both Theophilus (III.51) and Cellini, (1568) chapter 12, give vivid accounts of the process.

[69] The pre-baking of alum for use in such a flux is described in chapter 139. Neither the resulting alumina nor the calcothar of chap. 117 would be effective fluxes, for both are refractory powders. It may be that their presence in the soldering mixture served to control the spreading of the molten alloy, but since the alum in chapter 139 is said to dissolve the copper calcine it is more likely that the substance is misnamed and a flux such as borax is meant.

[70] Unless the heating were very restrained, this would give a pure gold powder—of no use as a solder for gold unless it were mixed with copper or copper oxide, which are not mentioned. It might do for friction-gilding, but it would be mainly valuable as a component of ink for chrysography. Note the comparable recipes for silver powder in chapters 121, and 132.

[71] Cf. chapter 132 for a similar recipe under a different title.

122.   Copper solder

Mix 1 pound of copper, 2 pounds of lead, melt the copper first, then put in the lead and mix into one.

122-A.   The recipe for tin solder

Mix 2 parts of tin and 1 of lead.[72]

122-B.   The recipe for a glue for stone

Take dried powder of white marble. Take one ounce of fish glue, and one ounce of ox glue, put them in water and cook them until they boil. Put in the marble powder and you make marble glue, stone glue.

122-C.   Another way

Take 2 ounces fish glue and 2 ounces of cheese glue, without marble powder as said above.

122-D.   A recipe for glue

Glue wood thus: Glue gold and silver separately with ox glue, or with fish glue as we taught above [in the chapter] on gilding.

123.   Glue from wood or bone

The glueing of wood in water: 1 oz. of fish glue, 1 oz. of ox glue 1 oz. [S] of fig-tree sap, 1 oz. of spurge sap: mix these together and cook them down in water. There is also a glue for carved woods: if it is wood on wood use one of the above-mentioned three [glues]. And [if bones are to be joined] on wood use 1 oz. of cheese glue mixed with 2 oz. of fish glue, cooked down into one. When the hot glue has been heated a little, glue the bones.

124.   Gold ore for smelting

We show you how gold can be made from the fattiness of an ore. When the ore has been discovered, make a pot that can hold 20 pounds of the ore. Then put it with the pot into the furnace and blow the fire from six o'clock to noon. Now afterwards you should put in the [roasted] fattiness of the ore 2 pounds of coral, 2 pounds of melted *ammoniacum*, copper calcine,[73] 2 pounds of Spanish salt, and as much white wax as is needed. 2 pounds of unguent and 1 pound of tartar: cooked down from every one of these pigments, each one permeating the other.[74] We have tested all these that you have read about. Because three [different kinds of] ores are involved in the cooking [i.e., smelting] of gold, we show you another

ore for cooking, but it will be more refractory than gold ore. For the man who wants to cook it, when he [recognizes it] by a dewlike odor in the very pot[75] where it is cooked, put with the first cooking half a pound of hard [S] pitch. Now in the second heating, he should put in some crushed glass, and in the third cooking, 2 pounds of tin to bring the ore to a sound result. And, when it is cooked, that which was mixed in the ore, turns to powder because it is assayed.[76]

125.   Silver ore

Prasinus is a green earth, from which ore silver flows. And this earth is engendered in rocky places where many ores of different colors are found. This rock when crushed has white veins, and when they are cooked, black stuff will emerge. It is tested like this. When it is broken up after it has been cooked, it shows colors as of silver inside; this stone is the one from which silver will emerge. Put this ore with cadmia from the upper part of the furnace onto the hearth of the smelting furnace[77] and fill up with coals; and so, *stratum super stratum* with wood and charcoal on top, smelt it as I said above. Melt it for a day and let it cool in the same place. Next take up the lump and break it up into tiny pieces, and put it back in the same furnace as before and with it [some] feminine lead: for a hundred-pound lump, 15 [pounds] of lead. Cook as before for 3 days. After this throw the lump out and break it up; put it in a furnace[78] and melt for 2 hours.

126.   The stone adamans

The stone *adamans* is engendered from cadmia and in the cooking of gold, in the first cooking of the lump. After the first cooking, while you are breaking up the lump (for the whole lump is easily broken up lightly) yet the [*adamans*] stones remain, some small, some large, which neither iron nor any other stone can overcome. This itself is stronger than all; it is vanquished by lead alone, and this is the power of lead.[79]

[72] Chapters 122-A through 122-D are not present in *P*, but are taken from *S*, folio 17v and 18.

[73] *Caucumarum*, *S*; *Calcum aurum*, *P*—supposedly both corruptions of *caucucecaumenon*.

[74] This is a complicated recipe for making a reducing alkaline flux like the simple black flux of the later assayers.

[75] Latin unclear. The smell given off on first roasting refractory ores containing sulphur and arsenic would serve to identify them.

[76] The roasting of the ore and its fusion with fluxes suggest that this chapter is a garbled version of methods of assaying gold ores, both easy and refractory ones. However, the lead which is usually used as a collecting agent is here replaced by copper and tin, neither of which would be desirable in the later cupelling operation.

[77] Follows *S*, *in catinam camini*. *P* reads *in canciacami*. Elsewhere in the manuscript, furnace is always *fornax*, the common word.

[78] *Calida vel in canicla S*, *calida vel in tanida*, *P*.

[79] *Adamans* was a hard stone, commonly diamond or corundum, neither of which seems appropriate here. The beginning of this recipe sounds a bit like the fusion of a sulphide ore to give slag, matte, and metal—the lead collecting the precious metal. But the end seems to describe the formation of a hard intermetallic compound, though not many such are soluble in lead. And what is cadmia doing? No zinc compounds are very hard. Or, is

Take female lead, soft and malleable, and melt it and throw in there the piece of *adamans* that you want to disintegrate; and heat up the lead over a slow fire; and, as it [the *adamans*] begins to disintegrate, immediately pick it up with tongs, and cover it with olive-oil soap, smoothly and very cleanly because it may be weak. For it is more fragile than glass, and softer than lead, since it may be melted in lead. Then take it out of the soap to disintegrate and the water will fall out and free it from the soap [*S*]. Then put as much as you wish carefully in a big fire and let it heat for 2 or 3 hours, until it is completely white-hot throughout. Afterwards take it out and wash it, and there will emerge *adamans* which fire does not overcome, nor does it shatter when struck, and it does not wear down when you work with it. By means of this you can copy [*S*] everything you want to work on.

127. *The purple dye from the murex*

Murex is engendered in every sea, more than in island places. It is a little shell, which has in it a place for blood, and the blood is a reddish purple: from this the purple dye is made. It is collected like this. Take the murex and collect the blood with the flesh and take some brine from the sea and put them together in a pot and leave it.[80]

128. *Yellowish-purple*

Take Alexandrian alum; grind it properly, and put it on a dish and pour boiling water over it; stir it for a time and let it settle. Afterwards strain off the hot water and agitate; then put in more hot water and agitate it; and place in [the alum solution] whatever you have to dye. Cover it and leave it for 2 days. After this stir it, let it settle[81] and leave it there 3 more days, and after this stir it around in the same way and leave it another 8 days and agitate it not more than twice a day. Then take it out and put in more alum. Then make another batch of dye and put it in, and next take clean urine from good wine and healthy men and take this urine and clarify it once, and afterwards put it in a copper cauldron; and take the same murex and wash it once lightly in water. After this grind it, and put it in thin cloths and wash it down in the urine in the cauldron. After this take some pig's blood and rub it [*S*] washing it also well in the same way. 1 pound of pig's blood to 3 oz. of murex.

About pig's blood. After this wash once a little pig's blood and rub it down; put it in the cauldron and make it boil a second time, and a third, in the same way, namely, 1 pound of the dye, 1 pound of murex with blood; i.e., 9 oz. of murex, and 3 of pig's blood.

129. *Making a bright purple from roses*

For making a bright purple from roses take the cookings of three cauldrons; into one put as much as you want of the concoction and the same amount of alum. Now if you want to dye it more cleanly, put into one [the second] pot just as much as in the first. Now there will also be a third dyeing in the same way.

130. *Yellow purple*

Now first a yellow dye is made; after this purple enters into the dye when it is made.

130-A. *The dispersion of gold*

Dispersion of gold. In a small crucible put pure gold mixed with quicksilver. Leaving it on the fire melt it, taking care that the quicksilver does not evaporate. Then take it out. Then triturate the true amount, mix the powder, and mix it again. Apply any quantity where you want it. [?][82]

131. *Gold paint, or a dispersion of gold*

Gold paint. Mix the powder of ground gold, [made] just as we said above, i.e. by the drying of quicksilver, 2 parts of gold powder and 1 part of verdigris with the composition; mix it again and make use of it as you wish.

132. *Silver paint, or a dispersion of silver*

Mix clean silver with quicksilver. After this put it on the fire and dry it with that quicksilver.[83] Then take the silver and grind it until it becomes powder: mix it with the composition, mix it again and make use of it where you wish.

132-A. *Another dispersion of silver* [*S*]

Take clean silver and mix it with quicksilver, as we said above. Then put it in a small cup, and put it on the fire until it throws off the quicksilver. Afterwards take 2 parts of silver and 1 part of verdigris

---

this a way of preparing some abrasive compound, which floats as a dust out of lead?

[80] Roosen-Runge, 1967: **2**: p. 25–30 has a good discussion of murex (*conchilium*).

[81] *L* reads *fac, quod iosu susu*, *S*, *iusum et susum*, *P*, *visum ÷ sursum*. In *L* the rest of this sentence and chapter became separated and appear without caption immediately following the chapter on *ficarin*, our No. 223.

[82] This chapter, in *S* & *L* only, is virtually incomprehensible transliterated Greek jargon. See note in Hedfors, 1932: p. 202. The phrase "mix it again" here and in chapters 131–132 translates *daufira*, which according to Svennung (1941: p. 95) is a condensation and latinisation of the Greek *d'au phura*, read as a noun by the scribe.

[83] This sentence not in *S* or *L*. It supposedly refers to the drying process of chap. 121 which is a near duplicate of this. The word translated as silver paint is *argirosantista*.

and mix it with some of the composition, mix it again, and make use of it.

133. *Emery stone*

The stone that is called emery, is rough and invincible, grinding everything: stones and gems are cut[84] with it.

134. *Lemnian earth*

The earth that is called Lemnian, which is a slightly purplish white, is engendered in rocky places; and you will recognize it by these signs. When it is wetted, it will boil and give off a strident sound. Now with alum it will dye everything green and purple except beryl and onyx.

There is an earth called black because it is dusky. Now it is engendered in Egypt, Africa, Havilah, and in Italy. It is engendered in moist places and in valleys. From it a rose dye is made; for when mixed with vinegar, and cooked, it will give this color; and afterwards it turns to scarlet.[85]

135 *Firestone*

The stone which is called the firestone, from which copper is cooked [S] is engendered everywhere. And there is another similar stone [which] when struck gives off rare great sparks; and it is reddish and fiery, having the color of copper. When it is put in the fire to be tested, it catches fire, and does not change its color. Collect it and grind it very fine; collect a medium-sized lump. Cover it with ox or goat dung and straw, and set it on fire for two days and nights, until the dung is consumed in the furnace. Now, you can cook this and copper and lead; and after it cools, the stone, which is now cooked, is collected. And in the first weighing, you weigh out 300 pounds for the first cooking. On it you put 18 basketfuls of charcoal [and] bundles of pitch pine. And when the masters of the work have arrived, set it all on fire and let it stand and cool, so that it does not run [S] like lead, iron, or other metal because it has become sluggish. While it is cooling, break it up into bits, melt it in the furnace, and work with it.[86]

136. *Split stone*

Split stone [slate] is engendered in Cappadocia, Asia, Iberia, and in Italy. It is dusky and strong.

When it is broken up small, you will find white veins in it; and, when it is burnt, it will become reddish. This is what the Alexandrians call "cadmia," because it melts glass. Now, it is engendered in high and windy places. And it is a stone that is easily split.

137. *How wax-marble is made from* gagatis.

The *gagatis*[87] stone is similar in color to orpiment, but it is not so green; and when it is broken, it gives off fire and splits into sheets. For this reason, the Alexandrians call it flat stone; and out of it wax-marble is made. If you crush it fine and put a pound of this stone, and 2 of ox glue,[88] and 5 pounds of water, and make it boil two or three times, stirring continually and mixing it together, wax-marble will result.

138. *The Thracian stone*

Thracian stone is engendered everywhere; now, it is green, split, dusky; when burnt, it will become white; it is put in cadmia in the cleaning of silver.

139. *How copper calcine is made*

Copper calcine[89] is made in this way. Make leaves out of very clean copper and put these leaves in an unused pot [S] with some ground natural sulphur; and spread out the leaves in a cooking-pot; then sprinkle sulphur on top, and again put leaves and sprinkle sulphur [S]; do so repeatedly until you have filled the pot. Then, place the pot in a glassworker's furnace, and cook for 3 days [4 days, S] and, when it has cooled, break it up very small. Take Asian alum, in amount corresponding to the sulphur composition. In the same way a pot [containing the alum] should be covered and plastered with potter's clay and placed [in the furnace] as in the former arrangement, and cooked for 6 days. And when it has been broken up, it should dissolve the copper calcine [i.e. act as a flux for it] to [make] a solder for gold.

140. *The recipe for electrum*

Electrum will be made in this way.[90] Put 2 parts of silver, a third of copper and a third of gold; in

---

[84] *Limantur*, literally filed. **S** says simply that emery sets everything on fire, **L** that it cuts all glass.

[85] These color changes suggest pH-sensitive organic dyes, not a mineral earth.

[86] This chapter seems to be an attempt to describe the roasting of pyrite prior to smelting it, or perhaps it refers to some kind of matte smelting. It is not inconceivable that cast iron was being produced—though certainly not from a copper ore. See the note to chapter 225.

[87] *Gagatis*. This word which was first used for a black combustible stone, perhaps jet, was later often confused with *achatis*, agate. Neither seems appropriate here. The "wax-marble" is a moldable artificial stone; *cf.* chapter 122-B.

[88] **P** reads *aurocollon* (=chrysocolla) for *taurocollon*!

[89] The recipe, which is one of those taken from Dioscorides, would produce copper oxide, or copper sulphide if air was completely excluded in the heating. Its use as gold solder is referred to in chapters 117 and 118, in the notes to which the metallurgical role of the calcined alum is discussed.

[90] This is a poor man's electrum, containing 50 per cent silver, 25 per cent copper, and 25 per cent gold (if one reads the silver as a $\frac{2}{3}$ part), or 75 per cent silver, 12.5 per cent of both gold and copper (if the *tertiam* is read as $\frac{1}{3}$). Electrum was initially a native alloy of gold containing enough silver to change its color (usually less than in this artificial alloy) and very little copper.

such a way that the gold and copper are of equal weight.

141. Gold solder for fistulas

Gold solder for fistulas from gold strip: take 1 ounce of *asticum*, 1 oz. of copper calcine, 1 oz. of *afronitrum*,[91] 1 oz. of olive-oil soap made without lime, 2 *solidi* of vitriol [*vitriolum*], half an ounce of vinegar, 1 oz. of water. Break them up and mix them, the copper separately and all the rest into one. Compound a little [of these together] as a solder to make, when heated, tubes from gold strip.

142. The recipe for making litharge from lead

Two kinds of litharge[92] are made, one from lead, and another from silver. Compound that which is made from lead as follows. Put rather soft feminine lead in a cooking pot and melt it well; then, when it is molten, continually skim the lead with a wooden pestle: at the same time, put in ashes [with live coals and skim and again put in ashes (*S*)] and do not stop skimming until you make it like powder; and afterwards wash it with water. Now, if you want it to be compacted and become thick, put it in a cooking pot or in a little cup [*L*] or into little tubes with oil; and when it is heated, it coagulates and when it cools, break off the tubes and it will come out dense [*L*].

143. Another recipe for making litharge from silver

You compound litharge from silver like this. Melt some [lead-bearing] silver and grind the scum that comes out of it with oil. It appears [on the surface of the molten metal] just as in the former recipe. However, because of the strength of the silver, it burns more strongly. Litharge from lead, on the other hand, before it becomes solid, enters with water into the glazing[93] of earthenware. But when it is striated, [*S*] it will be useful, wherever you want it.

144. Making gilded mosaic

Make a thick leaf of glass and put it [*S*] on a copper leaf, in such a way that when it is fired they will not stick together. After this put a gold leaf on top of the glass leaf and on the gold leaf put another very thin glass leaf; and put both in the furnace until the glass leaf begins to melt; and then remove it so that it cools. Next, rub its surface on a lead plate

dressed with emery until you have thinned the [glass] surface to bring out the color.[94]

145. Emery plates

Make a plate of lead. Take live emery, grind it well, and sprinkle the whole plate, rubbing glass down onto it at the same time, until the emery powder is embedded [*S*] into the plate. After this work whatever is necessary, using water.

146. To bring out the color of mosaic [by polishing]

Now, for bringing out the color, take a plate [of lead], and scrape it down. Then take some finely ground silver[-polishing (?)] earth, sprinkle the plate with it, and rub down the glass until the color is brought out well.

146-A. How to change copper into the color of gold [L]

Mix in a cup 2 parts of clean copper filings and 1 part of Asian alum, carefully pounded in a mortar and sieved. Put it on the coals, until it is melted, and let the alum be mixed with the copper. Now, first clean with urine whatever shape of pot that you want to make and pour the copper into it. In the first heating certainly and in the second, it keeps its color. In the third, it loses it if you have filed it, but [when] you beat it out, it retains it. If it breaks, it will be useless.

146-B. [The polishing of gems]

Each gem of the harder kind, such as the jacinth, the emerald, the almandine, the carbuncle, is rubbed with [powdered] emery stone on a lead plate, until it takes on the shape that the engraver wishes to give it. Then it is rubbed in a washing [i.e. levigated suspension] of the same powder with which it was first rubbed, until it is smooth.

Now, the brilliant polish is given in the same way: the jacinth, with the powder of calcined pyrite on a copper sheet, while the others are given their brilliant polish either with powder made from Cimolian earth or with that from an earthenware sherd of the kind that comes from antique pots. [The gems are held on the end of] a small pointed stick of aspen or alder wood.[95]

146-C. How gems of a softer nature are polished [S]

But gems of a softer nature (such as amethyst, rock-crystal, onyx, jasper, and beryl) are rubbed into shape with sandstone powder on lead. Then they

---

[91] *Afronitrum* for soldering is made as in chapter 221, but see also chapter 280. This whole chapter is repeated verbatim as chapter 218. *Asticum* is *araticum* or *arsaticum* in *S*; *et taritum* in *L*. Svennung (1941) thinks these may be misreadings of *eramenti*, an abbreviation of *eramenti*.

[92] Litharge is *litargirum* here and in chapter 143, elsewhere *spumae argenti*.

[93] *Bituminatio*. See note to chap. 108.

[94] This chapter is the one with which Muratori's transcript of the Lucca manuscript opens. It there bears the glamorous title *Compositiones ad tingendum musiva*, which for many years was treated as the title of the entire work.

[95] For the method of attachment, see chap. 146-F.

are rubbed down to smoothness in a washing of the same powder. They are given a brilliant polish in the powder of calcined pyrite on a copper sheet.

146-D. On [polishing] glass [S]

Glass, however, should be rubbed into shape on sandstone; then to smoothness with fine sandstone powder on lead. Then [while it is held] on a small pointed stick of wood, the work is completed by rubbing it on an antique sherd as on a whetstone, with water; finally it receives its brilliant polish on Cimolian earth, and this [too while it is held] on a small pointed stick of wood.[96]

146-E. How emery stone is prepared for polishing gems [S]

The emery stone is broken into a very fine powder, using a hammer on a hard anvil. And there should be a lead plate, which is fixed onto a wooden bench; and the powder of the [emery] stone is sprinkled on the plate and every kind of gem stone may be brought to shape on it. They are rubbed, with water, until they take on the shape that the engraver wishes to give them. Then the same powder is taken and washed; and the part that is the finest is put on another lead sheet, and any kind of gem stone is rubbed on it until it is perfectly smooth.

146-F. How unpolished gems ought to be held for polishing [S]

A piece of wood is taken, as thick as the little finger, as long as the width of a palm; and on its tip is placed hot pitch, mixed with ground-up tile—this mixture should have two parts of tile powder and a third one of pitch. After the mixture is heated [and placed on the stick], the gem stone that is to be polished should be applied so that it sticks to it.

146-G. Red Copper [and the etching and gilding of iron]

Red copper filings [S] are ground in a bronze[97] mortar with vinegar, salt, and alum to the consistency of honey. Some people use water instead of vinegar. Then the iron is well cleaned and gently heated and coated with this mixture and rubbed until it takes on the color of copper. Then it is washed off with water and rubbed; and [the iron] is gilded in the same way as copper or silver and heated to drive off the quick-

silver as usual. Then it should be rubbed with a [burnishing] tool so that it acquires brilliancy.[98]

146-H. Alum [(S) for etching iron]

Rounded alum, the salt that is called rock salt, blue vitriol, and some very sharp vinegar are ground in a bronze mortar; the cleaned iron is rubbed with these [materials] using some other kind of soft little point. And, when it has taken on the color of copper, it is wiped off and gilded, and then, after the quicksilver has evaporated, it should be cooled in water and rubbed with a tool that is very smooth and bright until it becomes brilliant.

147. The recipe for cadmia

1 pound of clean copper, 2 oz. of calcothar, 1 oz. of afronitrum, 1 oz. of sulphur. Put all these in a cup and melt them together and cook them until the copper and the calcothar are both burnt and what remains is released as cadmia.[99]

148. Quianus will be made like this

Mix and burn 1 part of copper, 1 part of lead, 1 oz. of ground natron, 1 oz. of calcothar, 1 oz. of afronitrum; mix with vinegar and put in the sun, let it dry, and grind it.[100]

---

[96] Cf. chap. 146-F.

[97] The word translated bronze is the genitive of aes which we have elsewhere translated as copper. In the lack of more precise terminology it was, of course, used for copper alloys also. Mortars hard enough to withstand pounding would surely have been made of bronze.

[98] See also chapters 219-A and 245, which give similar recipes, and chapters 291 and 292 which are almost verbatim copies of this and the following chapter. The treatment with the mixture described would provide a better "hold" for the gold by roughening the surface of the iron and depositing some copper by electrolytic replacement, for some copper would have been dissolved from the filings and the mortar in chapter 146-G and is added as copper sulphate in 146-H. Note that the corrosive solutions used here are chemically similar to those that later became important in the decorative etching of armor and in the first graphic etching. Both of these applications require a layer of wax, pitch, or linseed oil to protect the metal in the reserved portions of the design. Such decorative etching was used on iron swords of the La Tène period but it seems to have died out thereafter. All-over chemical attack remained in use to develop texture in Damascus and pattern-welded swords, but there is no more evidence of the use of stop-off coatings in Europe until early in the fifteenth century. Materials that would have been suitable are frequently mentioned in other connections in the Mappa. Etching was later used to reveal the structure of metals for microscopic examination—the very basis of the modern science of metals, see C. S. Smith, A History of Metallography (Chicago, 1960).
Nitric acid was the first mineral acid to be discovered. Initially called aqua martis in tribute to its use in etching iron armor, it may well have been produced by heating a mixture containing ferric sulphate and saltpeter intended for etching.

[99] This seems to be a recipe for producing zinc oxide from traces of zinc in copper—hardly a profitable procedure, even if the "copper" were brass.

[100] This is apparently an inorganic blue—mainly basic copper acetate, darkened if not entirely obscured by iron. Quianus was a common ingredient in the mixed pigment pandius (chap. 107-A n).

149. Anfinus[101]

Take soft lead, and melt it in an earthenware pot that is strong enough to withstand grinding. Take a pestle, and put coals with ashes over the lead. Before it cools, stir it smoothly and well with the pestle, until you thin out the lead and make it into fine [powder]. After this put it on a wooden platter, and wash it. Then compound it with sulphur in a new cooking pot, and cook for 3 days.

150. The recipe for a pandius

1 part of white lead, half a part of cinnabar; grind well in a marble mortar; and after it has been ground up, add some water in which fish glue is cooked, and a *pandius* pigment will result.[102]

151. Another kind

2 parts of verdigris, 1 part of cinnabar, 1 part of white lead, 1 part of *quianus*, 1 part of *lulax*.[103]

152. Another kind

1 oz. of *quianus*, 3 solidi of white lead, 1 part of natron, 1 part of calcothar.[104]

153. Another kind

1 part of alum, 1 part of native sulphur, 1 part of natron.

154. Making a green color in glass

Grind glass well, and put 3 oz. of clean copper filings to a pound of glass, and cook it for 3 days.

155. Another kind

When the glass has been ground well there should be added to one pound of it 2 oz. of copper filings and 1 oz. of Egyptian alum, and cook it for 3 days.

156. Making [glass of] a milky color

Put 3 oz. of tin to a pound of glass and cook it for 2 days.[105]

157. Making [glass of] blood color

Put 3 oz. of cinnabar to a pound of glass and cook it for 2 days.[106]

158. Making [glass of] a reddish color

Put 2 oz. of white lead to a pound of glass and cook for 6 days.

159. Making a purple color [on glass] without fire

Color thin glass pieces, mix and coat them with dragon's blood, and in this way a reddish color will result.

160. A pale apple green color [glass]

To a pound of glass, 2 oz. of Thespian earth and cook it for 3 days.

161. Red [glass]

To a pound of glass, 2 oz. of copper calcine.

162. Anthimis de damia

1 pound of *amor aquae*, 1 pound of naphtha, 3 pounds of native sulphur, 4 oz. of dry pitch [2 lb. balsam, 6 oz. *gagatis*, 4 oz. olive oil, 4 oz. resin (*S*)] 1½ pounds of milk of iron.[107] Grind all the dry ingredients well, and when very fine powder has been made, mix it into the liquids: and cook for an hour, and [a material to make] fire will result; however, not in accordance with the former strength but slightly less.

---

[101] In *P* the noun is hard to make out, although Phillipps in his printed transcript reads *Anfinus*. *S* reads *amphimissia* (fol. 45). *L* reads *antimis*, which Hedfords conjectured was a corruption of the Greek word *anthemis*, which is an herb like our camomile (from Dioscorides, Greek dict.). All seem equally incomprehensible. Whatever this recipe is, it has no relation to the incendiary *anthimis* in chapter 162. It describes a method of producing lead in a finely granular form by exploiting the extreme brittleness of the metal when a small amount of liquid remains between the crystal grains, and heating this with sulphur to make an artificial sulphide. Later assayers produced a lead powder by pouring the metal when on the point of solidification into a wooden box, and shaking it vigorously.

[102] This is similar to chapter 107-A.

[103] The correct medieval form of this word appears to be *lulax*, as in chapters 192 and 221-E. A transliterated Greek accusative form *lulacin* appears in chapters 151, 166, 168, 172, 173, 174, 176, 176-A, -B, -C, -D, -E and 231; a genitive *lulacis* in chap. 166; two misspelt accusatives in chap. 232 (P. *luzacin*: S. *lulazin*) and in chaps. 240 and 241 (P. *lulacerin*: S. *lulacerim*). Both *S.* (221-E) and *L.* (folio 229v) specifically identify *lulax* with indigo. Hedfors (1932: p. 103) notes the modern Greek *loulaki* (washing blue or indigo), and cites Diels to the effect that the word *lulax* derives from the Sanskrit *nīla* (n. Indigo) and *nīlā, nīlī* (f. indigo plant) via the Persian *nīlag, līlag, līlang*.

[104] *P* inverts the weights of *quianus* and white lead.

[105] See chapter 281 for a white potter's glaze also based on tin oxide.

[106] Neither this nor the succeeding recipe would be red unless some unmentioned ingredient, presumably copper or gold, was present.

[107] *Lac ferri.* Perhaps iron filings, though possibly the iron oxide that runs milklike off iron in the forge at a welding heat (1400°C or above). Berthelot thinks that this chapter describes a kind of varnish to apply colors to glass. To us, despite its environment, it seems more likely that it is an incendiary mixture, though it is the only one to appear also in *L*. It would have been more at home in the company of chapters 266–278 where many of the same ingredients appear. The collection of *amor aquae*, a natural petroleum, is described in chapter 276.

163. Olimpian stone

Olimpian stone is engendered in rocky places, and is duplex in color, black with white spots: when struck by the sun, it gives out fire, like sulphur.

164. The stone flavites

The stone *flavites* is engendered in black earth: and when it is struck by the sun it will become impregnated with *prasinus*. From it the green color *prasinus* is engendered.

165. The red stone

The red stone is engendered in diverse places; and from it also will be made the mortars in which gold is ground.

166. The recipe for lulax

[Take] flowers of parsley, flowers of clean flax, and a magma of violet of the two kinds above mentioned; that is, 1 part of the greater violet and 1 part of the lesser—now [make] such a magma not according to the recipe for azure [chap. 167], but only with water. Again such magmas may also be made of the greater blue lily—1 part. For use, both of the two magmas should be ground down into one and stored in a single glass jar. Make the magma of the lesser violet separately and make the magma of the greater blue lily separately. Then [take] 2 parts each of the parsley and the flax and 1 part of the lesser violet and 1 part of the greater one [and add] to one pound of magma 4 of frothed Egyptian alum, 2 solidi of the materials and 1 oz. of axle-grease soap without lime. Cook these a little, and grind 1 pound of deveined woad-leaves[108] and mix them with the cooked magma; and grind thoroughly until it becomes powder, and put it in the sun to dry. This is *lulax*, light in color approaching azure, and of a good color because it does not settle out since it is made of flowers.

167. The recipe for azure

Gather and lay aside flowers of *neulacis*, which in Greek is called *thapsia*,[109] though others call it chameleon plant. Then coat your hands with soap boiled without lime and rub the flowers between your hands and put them in a pot. After this coat your hands again with the soap and rub the same flowers for a rather long time and set them aside again; and

keep on doing this until the flowers are consumed. After the flowers have been consumed, collect the concoction and cover it carefully in a pot, in a hot place, until you notice that it has turned a blue color. Now when it is blue, cover it only with a cloth. Then take green leaves from a deveined woad plant, and cook them with defrothed urine, until the above-mentioned leaves are dissolved, and keep on cooking until the urine is consumed and the concoction thickens, and leave it to cool. Then take 3 pounds of *neulacis* flowers, and 2 pounds of the cooked woad leaves and half an oz. of cinnabar, mix them together, grind them cleanly, and when they are ground, let them stand covered in a mortar. Then put in a new cooking pot some oyster shells, carefully cleaned inside and out and washed free from dirt and mud, and cook them down to a powder;[110] and when they are cool, grind them carefully by themselves. Take 1 pound of the powder and some clean verdigris; put them in some other defrothed urine and grind for a while until it is turbid and the urine turns green, and mix some of this in the first mortar with the above-mentioned materials; grind well and store it in a new pot in the sun for a day. Afterwards cover the pot and lute it carefully and put it for a day in the upper part of a glassworker's furnace,[111] and azure will come out.

167-A. A diffuse azure which is called lively[112]

Soak flowers of *neulacis*, rubbed with soap, as we said above, in defrothed urine and cover the pot and put it in dung to disintegrate. In the same way soak woad leaves in defrothed urine; and after they have been kept in the dung and disintegrated, throw some of it into a mortar after removing all the veins from the leaves. Then take 1 pound of these woad leaves and 2 pounds of the *neulacis* flower and 2 oz. of poppy, mix and grind them together, adding a half oz. of cinnabar, a half oz. of verdigris, and a half oz. of defrothed urine, grind carefully, and put in a new cooking pot; cook it over a slow fire, until it is consumed and thickens, and there will result a slightly purplish azure.

167-B. An apple green azure [S]

Soak *neulacis* flowers in vinegar, place them in a covered pot, as said above, and put them in dung to disintegrate. In the same way carefully grind some deveined woad leaves, soak them in a new pot with vinegar and cover in dung, until the leaves disinte-

---

[108] The word for woad is here *guatum*, elsewhere in the manuscript *guattum*, *guuatum* and *uvatum*. The Classical Latin word was the same as that for glass, *vitrum*, explained by Webster as due to the blue color. The botanical term is *isatis* (from the Greek) *tinctoria*. The modern French *guède* retains the guttural, while the German *Waid* retains the *uv* forms of the Medieval Latin. See Hedfors, Roosen-Runge and Maigne D'Arnis *Lexikon*.

[109] *Thapsia*, which takes its name from the Greek island Thapsos, is a poisonous medicinal plant of the parsley family, used as a dye.

[110] Cf. chap. 174, where a kiln is specified for the "cooking" of oyster shells, which yield a purer lime than does limestone.

[111] Supposedly the annealing lehr, though this would certainly destroy any organic matter.

[112] Heading from *S*, *Lazurin difuton quod dicatur vivace*. Both the division into chapters and their headings from here through chapter 189 differ in *S* and *P*. We follow the former.

grate. Next take 1 pound of woad leaves, 1 oz. of lac cooked in defrothed urine, 1 pound of *neulacis* flower and 4 oz. of clean washed oyster-shell powder, as described above. Grind all these in a mortar and add to them half a pound of celandine plant concoction, which you have cooked in urine, and 1 oz. of saffron. When all this has been ground, cover it in a pot, and leave it for a day in dung, then take it out, dry it in the sun and use it.

168. *Sky-blue azure*

Take *neulacis* flower, rub it with soap, as we taught above, cover it in a pot and put it in dung. Do the same with deveined woad leaves and after some days, when they have rotted, take 1 oz. of the *neulacis* and the woad leaves and grind them fine in a mortar, adding 1 oz. of clean white lead, half an oz. of clean artificial *lulax*, half an oz. of cinnabar, and 4 oz. of urine defrothed with ground vitriol [*vitriolum*]: then add 10 pounds of urine, and after the vitriol has settled, grind enough urine in a mortar and leave the mixture to settle for 2 days. After this take 3 pounds of clean defrothed urine and 1 oz. of ground gall nut, mix them and let them soak for a day. Then take 1 pound of the stew, grind it well, and let it settle in the sun, and a sky-blue azure will result.

169. *An azure of a flesh-color* [S]

Take 1 oz. of the first kind of azure, 1 oz. of cinnabar and compound it as above.

170. *Onyx-colored azure* [S]

Compound as above 1 pound of clean, ground white lead, 1 oz. of azure, 1 oz. of cinnabar.

171. *Azure of a color like the eagle stone* [S]

Take 1 pound of *neulacis* flowers, coated with soap as was shown above, put them in dung, and 1 pound of deveined woad leaves with soap, as you[113] did above, and kneaded in dung. Afterwards, when they have been ground in a mortar, add to them 1 oz. of cinnabar, and from boiled down yellow-weed plant [S]—which should be boiled with defrothed urine until it is reduced to a $\frac{1}{3}$ part and thickens—of this thickness, when ground, 1 pound. Mix all these together and put them in the sun, and azure of an eagle-stone color[114] will result.

172. *A reddish color is composed of four substances* [S]

1 oz. of cinnabar, 1 oz. of *siricum*,[115] 1 oz. of cooked lac. Now lac is cooked like this: over a slow fire,

thoroughly cook finely ground lac in defrothed urine. Take 1 pound of this concoction and 1 solidus of *lulax* and after grinding them together leave them to settle and to dry in the sun.

173. *The recipe for purple* [*dye* (S)]

Grind well together 1 oz. of cinnabar, 2 solidi of *lulax*, and 1 solidus of white lead. Then dry in the sun.

174. *Another recipe* [S]

Mix and grind all together 1 oz. of the juice of pressed poppy flowers, half an oz. of cinnabar, 1 solidus of *lulax*: dry in the sun.

174-A. *Another recipe*

4 pounds of cinnabar [i.e.] vermilion, 1 pound of the earth vermilion that grows on the leaves of the Turkey oak,[116] 1 pound of the above-mentioned cooked lac, and 10 pounds of defrothed urine. Take both kinds of finely ground vermilion and put them, in a loose-textured fine linen cloth, into the urine in a cooking pot; wash the vermilion in the pot in which the urine has been cooked, and cook it and grind it again, and wash it in the urine in the pot. Keep on doing so until all the kermes is consumed. Then thoroughly cook this mixture and shake it: then take a clean, well-washed oyster shell, put in a well-covered pot and place it in a furnace until it powders, and then grind it fine. Put 3 pounds of this powder into the above-mentioned concoction and let it boil well down to a third. Then put it in the sun to thicken.

175. *Another recipe for vermilion* [S]

1 pound of vermilion, 1 pound of kermes—kermes grows, as was said above, on the leaves of the Turkey oak—1 oz. of cinnabar, and 1 oz. of the first azure. Mix them together; grind them carefully in a mortar and take 15 pounds of defrothed urine and cook in a new cooking pot, until the urine is reduced to half. Afterwards grind the [kermes] grains with cinnabar, pound and wash them in a linen cloth, just as they were contained in above, until it is consumed.

176. *Another recipe* [S]

Thoroughly grind half a pound of vermilion, 6 oz. of the other vermilion [kermes], 6 oz. of white lead, 6 oz. of *lulax* [S] and put these in a cooking pot with 10 pounds of defrothed urine, and putting them in a loose-textured linen cloth, wash down the kermes in

---

[113] Note second person singular—for the first time in the *P* manuscript.

[114] *Lazurin aetizonta*. *P* reads *melinizonta*.

[115] *Siricum*, made from lead or white lead (chap. 192-c) is minium or litharge.

[116] *Cerus*, the Turkey oak, is a shrubby evergreen tree of the Mediterranean region on which the kermes insect flourishes. "Earth vermilion," *vermiculum terrenum*, is identified as *coccarin* (kermes) in the next chapter. The insect's scales, which are a brilliant scarlet, are the oldest dyestuff of record. For a good discussion see Roosen-Runge, 1967: 2: pp. 40–51.

the urine, and again wash it until the kermes is spent. Boil it down until the urine is reduced to half and set it in the sun.

176-A.   A pandius [S]

1 pound of *lulax*, 1 pound of cinnabar, 1 pound of white lead, 2 oz. of *ficarin*: grind all these together and mix them with warm water; put them in the sun until they dry.

176-B.   Another pandius [S]

1 pound of *lulax*, 1 pound of prime cinnabar, 1 pound of azure, 1 pound of very clean ocher and 1 pound of *quianus*: grind all these together well and mix them with warm water, rub them and put them in the sun until they are dry.

176-C.   Again, another recipe [S]

3 oz. of *lulax*, 9 oz. of white lead.

176-D.   Again [S]

1 oz. of *lulax*, 1 oz. of *ficarin*, 1 oz. of *quianus*.

176-E.   A green pandius [S]

1 pound of *lulax*, 1 pound of *quianus*, 1 pound of white lead.

176-F.   Again, another recipe [S]

2 pounds of *quianus* and half a pound of white lead: mix and grind with enough defrothed urine; put in the sun.

176-G.   Again, another way [S]

1 pound of *quianus*, 1 pound of *ficarin*, 2 pounds of ocher: grind them all together, mix with defrothed urine and put in the sun.

176-H.   Again, another way [S]

1 pound of *quianus*, 1 oz. of ground copper calcine, 1 oz. of *ficarin*, 1 oz. of ocher: grind these all together, mix with defrothed urine and put in the sun.

176-I.   The first pandius of a cinnabar color [S]

1 pound of cinnabar, 1 pound of boiled-down cooked yellow weed, 1 pound of clear saffron-yellow millet, 2 pounds of *ficarin*, 1 pound of *quianus*: grind them all together, mix them with defrothed urine, and dry in the sun.

177.   Another pandius of a cinnabar color [S]

6 oz. of cinnabar, 6 oz. of the cooked lac broth and 6 oz. of saffron: grind them all together and put the mixture in a glass pot in the sun for a day, until it dries, and take it inside at night.

178.   Again, another recipe [S]

Grind and mix together 1 oz. of cinnabar, 1 oz. of the three ingredients of ink. Put it aside in a glass pot, place in the sun, and take it in at night; and keep on doing so until it dries.

179.   Again, a pandius [S]

Grind well in a mortar 2 oz. of cinnabar and 1 oz. of white lead. Mix them with defrothed urine and grind properly; store them in a glass pot and cover it in dung for many days.

180.   Again, a green pandius [S]

Grind and mix 2 oz. of green earth and 1 oz. of cinnabar, and store them any way you like, after first mixing them with defrothed urine.

181.   Again, a green pandius [S]

1 pound of green earth, 1 oz. of cinnabar, 2 solidi of white lead. Grind these in a mortar with defrothed urine, and store them in a glass pot, and put them in the sun to dry.

182.   Again, a green pandius [S]

1 pound of green earth, 1 oz. of ocher, 1 oz. of cinnabar; grind and mix them all together with defrothed urine, put in an earthenware pot, and cover in dung for 20 days.

183.   An Ocher-colored pandius

1 pound of clean ocher, 1 oz. of cinnabar, 3 solidi of *ficarin*: grind them all in a mortar, mix with defrothed urine and store them in a glass pot and put them in the sun, until they dry.

184.   A purple pandius composed of four ingredients [S]

Lulax, *quianus*, cinnabar, lac, in equal weights: grind and mix, put them in a glass pot, and put in the sun until they dry out.

185.   Again, a purple pandius [S]

1 pound of the broth made by boiling down murex;[117] [1 lb. cinnabar (S)], 1 oz. of clean *siricum*: grind them all together and mix them with a little urine, put them in a glass pot and dry in the sun.

---

[117] *Cf.* chap. 127.

186. *Again, a purple* pandius [*S*]

Murex broth, lac broth, 1 oz. of each: first grind cinnabar, 1 oz. and after this mix in the murex broth and the lac broth and set aside in a glass pot in the sun until it dries.

187. *Again, a pale purple* pandius [*S*]

1 oz. of the murex broth, 1 oz. of cinnabar, 1 oz. of saffron, 4 oz. of yellow-weed broth: cook them down with urine, all mixed together, to a sixth of their weight [*S*].

188. *Again, a* pandius [*S*]

1 oz. of cinnabar, 1 oz. of murex broth, 1 oz. of cooked madder, and as much of cooked *finiscus*:[118] first grind the cinnabar by itself, then mix them all together and put them in a glass pot, just as in the other cases.

189. *A* pandius [*S*]

Take madder broth, and add 3 oz. of gallnut and grind it properly; take 2 pounds of madder broth and put it in a glass pot with the ground gallnut and leave to soak for 2 days: after this, strain it and add 1 oz. of calcothar and 2 *solidi* of cinnabar [and] grind them both, and put them with the above-mentioned things, and cook down until it is reduced to a third.

190. *The recipe for green ink*

Take ripe seeds of the shrub *caprifolium*, that is in English gatetriu,[119] and grind them together well in a mortar; afterwards, let them boil thoroughly in wine, at the same time adding to the concoction iron that has rusted. This is a brilliant green ink. If you want to make cloth or leather green, smear some of this on it with a paintbrush. Now, if you want it to be black, add vitriol to this composition in the usual [way]. But, if you want to prevent this or any other ink from running, put some gum of hawthorn or holm oak into the concoction and cook them together.

---

[118] Perhaps a corruption of *faenisicia*, mown hay.

[119] Phillipps remarks in the introduction to his edition of *P*, "... In chapter cxc. ... the shrub 'caprifolium' is translated 'goat tree.' This is a singular circumstance, and seems to me to indicate, as I said before, that the author or the transcriber was an Englishman, for had he been of any other nation he would most naturally have translated it by the language of his own country. Moreover, in the very next chapter, he mentions the herb 'greningpert,' a corruption, I suspect, of 'greningwert,' the Saxon Þ being easily mistaken for a *p*; (and we know that several English names of herbs end in *wort*, as St. John's Wort, &c.) which I consider an additional mark of his being an English author." This comment, of course, should be applied only to the authorship of this chapter and the following one, neither of which appear in the earlier manuscripts. On medieval ink see Roosen-Runge (1972).

191. *To temper green*

Take the plant that is called greningwort,[120] and boil it well with beer or wine, until the beer turns yellow from the plant. Afterwards strain it; then mill powdered Byzantine green with the beer, and put in as much beer as is sufficient. Afterwards let it stand in a basin or a copper pot, to mature in the sun.

191-A. [*Cutting and polishing gems*]

Have a copper strip and fine emery powder, and, when you want to cut a stone, wet your saw [i.e., the copper strip] in the middle with a little saliva, put powder underneath, and holding the saw firmly or closely, apply it to cut the stone.

Polish [gems] in this way. Put emery powder on a lead plate, and wet it slightly with saliva. Polish [the gem] gradually, watching often to see that the powder is not used up. When the gem is cut [i.e., rough polished], bring out the color in this way. Pound some pebbles that have been calcined until they are reduced to powder; dry this in the daylight, and make a very fine powder from it; alternatively make a fine powder of light pumice-stone, or of calcined rock-crystal. Now, stretch a strip of leather over a plate and with the powder and your saliva bring out the color of the stone by rubbing it on the leather strip. If you do not have pumice-stone, make an equally potent powder from the ancient brick tile from which salvers used to be made, or you may bring out the color without harm with copper wire.[121]

191-B. Quianon ualtalasion

It is engendered in humid places, for it is engendered out of the dew in the summertime. It is collected as follows: pick it up and collect it and lay it in the sun until it is dry. After it is dried out, grind it well. Then take the greater sea snail, wash it well, grind it; take 30 pounds of this snail, 2 pounds of *ualtalasion*, 10 pounds of soap, 3 oz. of azure and a piece of lightly-cooked olive oil soap. Mix these all together, grind them and set them aside in a new earthenware pot. Cover them in dung and leave them for 60 days.[122]

192. Quianus, *then is engendered as follows.*

Before mixing the ingredients, weigh them, then make a good mixture, grind them well on marble and

---

[120] See previous footnote.

[121] *Cupero filo.* Perhaps a corruption of *coperosa*, copperas, which, when calcined, yields that excellent polishing powder, rouge.

[122] In *S* this chapter follows immediately after the material of chap. 189. *P* spells *ualtalasion* with a double *u*, and bears the chapter heading that we, following *S*, assign to 192.

mix them well in amounts appropriate to your operation.

192-A. *Again, another green* pandius [*S*]

1 pound of *quianus* and 1 oz. of white lead, both ground and mixed with defrothed urine.

192-B. *Again, a* pandius

1 pound of *quianus* and 1 oz. of cinnabar—grind and mix these with defrothed urine.

192-C. [*A list of things that painters use*]

We list all those things made from flowers of land or sea, also from plants; we list in this way their qualities or their uses on walls, woodwork, linen cloth, and even on skins. We mention all the things on which painters work and the techniques which they all use: the man who lays a simple coating of glue on a wall; one mixed with wax on woodwork; on simple wood, glue mixed with plaster; on a cloth however colors mixed with wax; on hides plaster[123] mixed with glue [is used].

192-D. *On the kinds of minerals, herbs, woods, stones, and also on fungi* (*L*) *salt, natron,* afronitrum, *oils, pitch, resin and sulphur* [*S*]

The first ore is that from which gold comes. It is a ruddy earth: it is almond-shape, reddish, on account of the adjacent earth. There is also another like it, which loses color when it is burnt and is not sandy like the former. This earth is engendered in sunny places. And such is ore of gold.

Now the ore of silver is green.

And the ore of copper is a green rock, and when you strike it with a firestone,[124] it emits fire.

And brass [ore] is an apple-green rock, and emits fire in the same way. The ore is a stone that is jet colored.

And lead is a dusky earth; and the stone which is found in it is green.

Sand is the mineral from which glass comes; now there is also a stone and it is of a glassy color.

Vitriol [*vitriolum*] becomes an earth. Ochers are grown where there are drops in springtime. They are collected and cooked down. And this earth becomes calcothar [when calcined]; but the earth that is dry is vitriol.

Alum ore is an efflorescent earth.

Eritarin [*S*; *eitarin*, *P*.] is a white earth, easy to break up.

Sulphur is engendered in the earth, and the very place burns. From sulphur-earth mixed with oil, a concoction is cooked.

Natron is a salt which is engendered in the earth. It will turn into sheets and in time become full of holes.

Salt schist is engendered in the same way.

Afronitrum, however, is engendered in the place of natron,[125] before it congeals; also another kind is made from natron; and the chief kind is a foam, white like snow. Now, when it is compounded, it is at first dusky, yet it has the same property.

Sulphurous earth is engendered in the same place as is sulphur. [Hematite is found near the place where sulphur is found (*S*)]

Quicksilver is engendered from an earth, and another kind from silver ore in the smelting process.[126]

The ore of orpiment is earthy.

Calcitis [*S*] is a natural clod of earth which is found on the island of Cyprus in mines, of a pale golden color; it has within it cleft veins like fissile alum, and they shine like stars.

Prasinus is an earth that is mineralizing.

Lulax is made from an earth and plants.

Azure is compounded.

Quianus [*Cyanus* (*S*)] is compounded.

Ficarin is compounded.

Verdigris is an efflorescence of copper.

White lead is an efflorescence of lead.

Ochery earth is a *pandius*—you can color everything with it, compound everything with it.

Copper calcine is made from copper.

Cinnabar is made from quicksilver.

Siricum [*Sciricum* (*S*)] is made from white lead; it is also made from lead.[127]

193. *Of plants, earths and woods*

Chrisocollon is a tree that is not tall and whose interior wood is apple-green. The bark and fruit of a nut tree: bark of Cilician [*Olicine* (*S*), *Cilicine* (*L*)]; bark of ash [*melia*] trees; bark of elm trees; and bark of the poplar tree.[128]

All these are dyes: madder from the forest is weld [*S*]. *Monoclosus* is an acorn gall. (The *tamus* is woodlike and resplendent,[129] *Titimalin* is a plant [*S*]). *Drantalasis* is the plant double bugloss.

Every kind of resin is cooked from pine and fir. When pitch-pine is cooked, pitch oil comes out first; and cedar pitch is cooked from cedar wood. (Pitch

---

123 *unctio*, supposedly lime plaster or gypsum in a kind of gesso.

124 *pyregliolo*, **S**, *pirepolo*, **P**, *pirebolo*, **L**.

125 The Lucca manuscript reads *vitri*, glass, instead of *nitri*, natron: Just possibly sandiver is meant. See note to chap. 1.

126 "[Hydrargyrum] is also found in ye place where silver is melted standing together by drops on ye roofs"—(Dioscorides, Goodyear trans., Book V para. 110). Supposedly the mercury came from cinnabar in the silver ore, for there is no evidence for the use of mercury in ore treatment before the sixteenth century.

127 *Siricum* (*cf.* chap. 172) is therefore probably litharge or minium (or an intermediate color between the two) made from white lead as in chap. vii, or by the direct oxidation of molten lead as in chap. 142.

128 Identification of these trees uncertain.

129 Phrase obscure, but see Hedfors, pp. 121–123.

is also engendered from fir [*sappinus*] and a fir pitch from the silver fir [*S*]). Mastic [*mastice*] is engendered from the mastic tree [*lentiscus*]. Hornbeam [*Zigea, zizipha*] is a tree the fruit of which is the jujube. Gum from the maple tree. A second gum, from the almond tree. Olive oil comes from the olive. Linseed oil comes from flaxseed. Mastic [*lentiscus*] oil from the mastic tree. Coral[130] from the sea. The murex from the sea. Salt from the sea.

194. [*The assaying of gold-silver alloys by weighing in air and water*]

Any pure gold, whatever its weight, is heavier than any equally pure silver of identical weight [when both are weighed in water] by a 24th plus a 240th part of itself; and it can be assayed as follows: If a pound of the purest gold is compared on the balance under water with the same weight of equally pure silver, the gold will be found heavier than the silver, or the silver lighter than the gold, by 11 pennyweight, i.e., by a 24th plus a 240th part of itself. Therefore, if you find any shaped [goldsmith's] work which you believe to be alloyed with silver and you want to know how much gold or how much silver is contained in it, take some silver or gold and, checking and examining the weight, make a lump of whichever metal you wish equal to the weight of the suspected work [*S*], and place both of these, namely the work and the lump on the [two] pans of the balance and immerse them in water.[131] If the lump that you made was of silver, the work will outweigh it; if it was of gold, the object will rise and the gold sink. Now, this happens in such a way that the silver rises by as many parts [i.e., fractions of the whole difference between gold and silver] as the gold sinks. For, whatever is in the work [when weighed] beneath water beyond its usual weight [in air] belongs to the gold because of its density, whereas whatever lightness there is must be attributed to the silver because of its rarity. To make this more easily observed, you should notice carefully that in the density of gold, as in the rarity of silver, 11 pennyweight signifies a pound, just as was stated at the beginning of this chapter.[132]

---

[130] Follows *L, corallum. S* reads *colaliam, P, collium.*

[131] In his transcription Phillipps omitted a sentence from the MS: (*Si argentea fuerit*) *massa quam fecisti, opus praeponderabit. Si aurea fuerit (allevato opere....*). Notice the second person singular again.

[132] More simply stated, the fraction of silver in the alloy is $240/11[(w_{Au} - w)/W]$, where W and w are the weights of the alloy in air and in water respectively, and w_{Au} is the weight in water of a mass of pure gold that, in air, equals W. The method requires the immersion of both balance pans in the water. The computation is correct only with pure alloys of gold and silver, and it is assumed that there is no change of volume on alloying. The make-up weights used under water should be silver. The variable effects of capillarity on the suspension strings at the water surface are ignored. This method correctly gives the fraction of gold by weight: other early methods often give volume

194-A. [*The ratio of weights of wax and metals for use in the foundry*]

The weights corresponding to one ounce of wax are: Tin, 7 oz. 17 pennyweight; white copper, 8 oz.

fractions, the difference between the two, which is large, being overlooked.

This account appears in nearly identical form in several other tenth-century and later sources. In Paris, BN lat. 8680A, 10v. 13c, it is added on the margin as a possible proof of Proposition IV of the Pseudo-Archimedean *De ponderibus Archimenidis* (or, *De insidentibus in humidum*). It is also included in a complex geometrical-metrological manuscript at Munich, Staatsbibl., cod. 14836, 137r. 11c, from which it was published by M. Curtze (1895). In the version included in the tenth-century manuscript in Paris, BN lat. 12292, and in the twelfth-century manuscripts of Eraclius the difference in water loss between silver and gold is stated as 12 pennyweight per pound, i.e., 1/20 which is less accurate than our figure. Using the modern values for the specific gravities of gold and silver (19.3 and 10.5 respectively), the fractional differential loss in weight of silver versus gold in water should be $1/\rho_{Ag} - 1/\rho_{Au} = 0.0434$ or 10.4 dwt/lb. However, if the weights used to restore balance were immersed alongside the alloy sample and were also of silver the loss would appear to be larger by the fraction $\rho_{Ag}/(\rho_{Ag} - 1)$ or 1.105, making 11.5 dwt differential loss per pound of silver. The stated value, 11 dwt/lb is remarkably close to both. It may have been the realization of the necessity of a correction for the immersion of weights that gave rise to the curious expression in the opening sentence "a 24th part plus a 240th part."

The method is a development of the Archimedean scheme for detecting adulteration of gold, first described in Vitruvius *De Architectura*. The poem *Carmen de ponderibus*, probably by the Latin etymologist Priscian and dating from about 500 A.D., describes a method based on the same principle as the present one but using a balance with movable fulcrum. The standard fractional loss on immersion is there given as 3 drachma per lb. of gold, i.e., 1/25. There follows a non-hydrostatic method involving the production of pieces of alloy, silver, and gold having equal volumes. Another method required the laborious (and inexact) production, via a wax intermediate stage, of a pure silver duplicate in shape and volume of the object, and weighing it. When, later, the ratio of wax to metal weights was given as in our following chapter, this was for the foundryman's benefit not the assayer's.

For the early history of the use of Archimedes' principle in assaying, see C. Thurot, (1868 and 1869); M. Berthelot (1893: 1: pp. 167–178) and especially Marshall Clagett (1959: pps. 64–69, 84–99, 114–124, 131–134). Clagett gives the text with English translations of the important sources including *Carmen*, BN 12292, and Sélestat. The work of the Muslim scientists al-Biruni and al-Khazini (early twelfth century) who determined the relative specific gravities of seven metals and a few gems was summarized by A. Mieli (1939) based upon the pioneering studies by E. Wiedemann.

The density assay in one form or another is mentioned in most books on assaying published in the sixteenth to nineteenth centuries—see especially Lazarus Ercker's *Beschreibung allerfürnemisten mineralischen Ertzt und Berckwercksarten . . .* (Prague, 1574) and the extensive notes added by J. H. Cardalucius to the 1672 edition. It was a popular problem in elementary textbooks of mathematics and gave rise to some early nomographs as well as to many specially graduated balances and hydrometers. Both Galileo and Boyle described such gadgets. Although it was too inaccurate for wide use with the precious metals, the density assay was used as an official test by the pewterer's guilds to check the adulteration of tin with cheaper, heavier, lead. They did not use hydrostatics but simply weighed castings made in a standard bullet mold against alloys of legal composition.

16 pennyweight; Cyprian copper, 9 oz. 3 dwt.; [brass 9 oz. 3 dwt.;] silver, 10 oz. 12 dwt.; lead, 1 oz. [i.e. lb.] 6 dwt.; gold 19 oz. 9 dwt.

And so if there is one pound of wax, put in of tin 7 lb. 10 oz. 4 dwt. You should put that in because for as many ounces as there are of the wax there should be [for each] 7 oz. 19 [sic] dwt. of tin, and so if there is a pound of wax, i.e., twelve ounces, put in twelve times 7 oz. of tin which makes 7 lbs., and twelve times 17 dwt., which makes 204 dwt or 10 oz. 4 dwt. Of white copper, if there is a pound of wax, you must take 8 lb. plus 12 times 16, which corresponds to 192 dwt., making 9 oz. 12 dwt.

Of Cyprian copper, for a pound of wax, take 9 lbs. 12 dwt. [sic]. Of brass, for a pound of wax, take 9 lbs. plus twelve times 3 dwt., which makes 1 oz. 16 dwt. On the other hand, with silver, to a pound of wax, 10 lb. plus twelve times 12 dwt.

In the same way to a pound of wax, put in 12 lb. of lead plus twelve times 6 dwt. When casting gold, on the other hand, for a pound of wax take of gold 19 lb. plus twelve times 8 dwt., which makes 3 oz. 13 dwt. [sic].[133]

[133] This chapter, missing in *P*, is translated from Sélestat (fol. 40, 40*v*). There is an additional version of the first paragraph only, with the correct weight for lead, on fol. 212*v*. The title is taken from BN 12292, cited by Berthelot (1893: p. 175), *De mensura cerae et metalli in operibus fusilibus*. Theobald (1933: p. 304) cites another manuscript of the tenth century with the same data in it, and it also appears in Munich codex 14836, from which it was published by Curtze (1895: p. 139). It is the earliest quantitative collection of any of the physical properties of matter. The values are tabulated below.

WEIGHTS OF METALS AND ALLOYS

Metal	Weight equivalent to one ounce of wax		Weight equivalent to one pound of wax			Computed density based on wax =0.96	Density (from *Metals Handbook* Vol. 1, 1966)
	oz.	dwt.	lb.	oz.	dwt.		
Tin	7	17	7	10	4	7.53	7.30
White copper	8	16	8	9	12	8.45	(8.8?)
Cyprian copper	9	3[1]	9	0	12	8.78	8.96
Brass	[9	3][2]	9	1	16	8.78	(8.5)
Silver	10	12	10	7	4	10.18	10.49
Lead	12	6	12	3	12	11.81	11.34
Gold	19	9[3]	19	3	13	18.68	19.32

Notes:

1. If computed backwards from the stated weight per lb. of copper this would be 9 oz. 1 dwt. ($\rho = 8.70$). There has been some confusion between copper and brass at some point. The identity of white copper is questionable—it could be a high-tin bronze or one of the white Cu-As alloys that were enjoyed by experimentalists but not by practical foundrymen.

2. The equivalent of brass per ounce of wax is not listed. This number given is computed from the weight per pound.

3. The Froumundus manuscript cited by Theobald reads 19 oz. 8 dwt.; the stated value per pound of gold corresponds to 19 oz. $6\frac{1}{12}$ dwt. per ounce of wax ($\rho = 18.58$). The pound and ounce equivalents for gold do not agree in either manuscript.

The weight system is 1 pound = 12 ounces = 240 pennyweights. The modern density figures in the last column of the table are

195. *The recipe for niello for use on gold*

Take 2 parts of *almenbuz* (i.e., silver)[134] and a third [part] of copper, and as much again and a bit more of *alquibriz* (i.e., sulphur). [Withholding the sulphur,] put them in a crucible in the furnace to be roasted; and then gradually mix in the above-mentioned *alquibriz*. When it is well roasted and mixed, take it out and pour it into an ingot mold or wherever you wish, and hammer it while it is still hot so that it is thinned out. Then leave it to cool. Afterwards thoroughly break it up very fine on an anvil with a pounder, i.e., a little hammer, until it is made into powder, and put it in a shell. Afterwards temper some *atincar* (i.e., borax) with water, and with this temper the niello, and place it where you wish [on engraved portions of the gold work to be decorated]. Sprinkle soda [*natronum*] powder on top, and put it on coals until the niello runs well. In those places where you do not want it to run, put some very fine chalk tempered [with water]. When this is done, take it out of the furnace to cool, and rub[135] it with a maple polishing stick. Heat it slightly but often on the fire, continuing until it is in good shape. Afterwards scrape the niello right down to the [level of the surface of the] *almenbuz* and polish it again as you know full well, and leave it.

those for pure solid elements. They are not adjusted for the fact that contraction of a casting during and after solidification gives it, when cold, a smaller volume than its mould cavity. The modern pattern maker allows about 3 to 4.5 per cent by volume for this. Castings of both wax and metal are liable to contain porosity which would upset the ratio of their weights.

[134] Note the use of Arabic words in this and the next eight chapters. The Latin identifications (which we translate in parentheses) are interlineated in the manuscript in a different hand. The word *almenbuz* clearly means silver as denoted by the interpolation, but it does not resemble the usual Arabic word of silver, *fiḍḍa*. The others are straightforward transcriptions, as follows:

alquibriz = *al-kibrīt* = sulphur
atincar = *al-tinkār* or *attinkār* = borax
arrazgaz = *al-raṣāṣ* or *arraṣāṣ* = lead
alcazir = *al-qaṣdīr* = tin
natronum = *al-naṭrūn* or *annaṭrūn* = soda (Not to be confused with Latin *nitrum*, natron. See footnote 13, chap. 1)

These chapters are not in *S*, which was compiled in the tenth century, before the influence of Arab science had begun to be felt in Europe. That they came via Spain is indicated by the use of '*z*' for '*s*'.

Note that this niello for use on gold is 2 Ag, 1 Cu, 3+ S, and that in the next chapter for use on silver is 1 Ag, 1 Cu, 2 S. There is no lead, although the niello of chapters 56 and 206, as well as that in Theophilus (III.28) and most later recipes, includes it.

[135] The intent here seems to be to eliminate cavities by compressing the niello when it is hot enough to be plastic, but not molten. However, the same word, *lipso*, appears later in this paragraph and elsewhere where simple abrasive polishing is clearly meant.

196. Again, niello for use on almenbuz [*silver*]

Take *almenbuz* and the same amount of copper and as much *alquibriz* as the *almenbuz* and copper together weigh; and do as was said above in the case of gold.

197. *Again, as above, so that the gilding may have color*

Take some urine and a little vinegar, and some well-ground garlic, and as much *alquibriz* as you think right, and mix them together in a copper shell; and put in there 2 little sheets, one of copper and the other of *arrazgas* (i.e., lead) and when it is boiling hot, take them out and dip them in cold water, but in such a way that they do not touch the bottom; immerse and remove them repeatedly until they have a good color.[136]

198. *To soften gold, do as follows*

Put gold in a crucible in the furnace and mix with it *alquibriz* and *tincar*. Melt them together, and make an ingot and put it in salt, then in water. Then work it.[137]

199. *Again, if you want to lay gold on a skin*

If you want to lay gold on a skin, first put white of egg on 2 or 3 times; if[138] *almenbuz*, 4 times; if[138] *alcazir* (i.e., tin), 8 times.

200. *If you want to color* almenbuz

If you want to color *almenbuz*, take vinegar and salt and mix them together, and then heat the *almenbuz* and quench it in them. Afterwards take ground charcoal and [with it] rub [the *almenbuz*] using a cloth or a bristle brush.

201. *If you want to join copper or brass*

Take 2 parts of copper, and a third of tin;[139] melt them together in the furnace, and mix them well; take it out of there to cool, and from it make very fine powder on a [flat piece of] iron or hard stone, and mix the powder with olive oil, making it neither too thin nor too thick. With this coat the copper or brass joint, and sprinkle over it powdered soda [*natronum*] (i.e., *alatronum* [*sic*]) and put it on the fire to heat, and rub with a little forked tool so that it joins well.

202. *The joining together of brass*

1 pennyweight by weight of soda, burnt cream of tartar[140] as much as you think right, 1 pennyweight by weight of borax: compound these with water, and coat the brass with it; afterwards sprinkle over it roasted tin powder;[141] afterwards heat it in a furnace beneath the coals, as in the case of [soldering] gold, until it is well joined.

203. *Joining tin together*

1 part of soap, 1 part of pine resin, 1 part of soda and some borax. Coat the tin with this and heat lightly, as you know how, until it joins together, and quench it while still hot in water.

204. *An easy way of gilding*

Take leaves of tin, dip them in vinegar and alum, and glue them together with glue made from parchment. Then take saffron and pure (i.e., clear and transparent) glue, drench them both in water with vinegar and cook them with filings [parchment scrapings?] over a slow fire. When the glue flows, coat the tin leaves [with it] and they will appear golden to you.

205. *For a tin solder*

Mix together two parts of axle-grease with a third part of resin and an equal amount of tin filings. Heat these lightly at the fire and you will be able to solder.[142]

206. *Painting black on a gilt vessel, so that you think it has been inlaid*

Melt equal parts of red copper and lead, and sprinkle native sulphur on, and when you have melted it, let it harden; put it in a mortar, grind it, add vinegar, and make the ink with which writing is done of the [proper] consistency. Write whatever you wish on gold or silver [vessels]. When it has hardened, heat it, and it will be inlaid.[143]

Now it may be melted like this: cut a cavity in a piece of charcoal, and put silver and copper in it; melt them, and when they become liquid mix in lead, then sulphur; and when you have mixed it, pour it out and do as was said above.

207. *Gilding by the application of [dyed tin] leaf*

Leaves of tin should be made; and they should be made like this. Melt the tin well, and slowly pour it

---

[136] This seems to be a recipe, not for gilding, but for producing a sulfide patina on copper and lead. The possibility of electrolytic action might be noted.

[137] This would soften the gold by removing silver and other metallic impurities in the form of sulfides.

[138] Text reads *in*, on, in both places. More adhesive seems to be needed with the stiffer foils.

[139] This solder—67 copper, 33 tin—is brittle enough to be easily powdered but not too brittle for use. It melts at about 740°C.

[140] *Cream vini assam.* The word *cream*, otherwise unknown, is curiously coincident with the English term used for purified crystalline tartar. The burnt product would be mainly potassium carbonate.

[141] "Roasted tin powder," *pulveris stanni assi*, is supposedly tin oxide.

[142] Chap. 261 is a duplicate of this.

[143] The same recipe, though with added silver in the first sentence, was given in chapter 56.

out onto a marble slab and make thin leaves of it. Alternatively make them with a hammer. Apply them in the same way as gold leaf. Cook down swallow-wort and take 3 ounces of this concoction, 3 pounds of saffron, and 1 pound of orpiment. [Coat the tin leaf with this.][144]

208. Dyeing tin leaf

Take 1 ounce of clean saffron, 2 ounces of the best split orpiment [S], add half [an ounce] of gum, and half an ounce of linseed-oil. Mix in rain or fresh water, mix and boil them together. Mix the preparation together, grinding it well, and taking it up with a sponge, coat the leaf with it. When it has dried, coat it a second time, and when it is dry, rub it with an onyx until it shines brilliantly.

209. Making proven gold[145]

Melt together 4 parts of copper, 1 part of silver, and add 4 parts of unburnt orpiment, i.e., crude *eureos* and when you have heated it, let it cool and put it in a pan—coat the pan with potters clay—and roast it, until it becomes a cherry red; take it out and melt it, and you will find silver. And if you roast it a great deal, it becomes *elidrium*; if you add 1 part of gold, it becomes the best gold.

210. [A solder from silver]

For a solder from silver, weigh out 2 pennyweight of silver, 2 of copper, and one *medalla* [½ pennyweight ?] of tin.

211. A solder for poor silver

Take good silver, weighing 3 pennyweight, and 1 obol of tin. For soldering good silver, take half an obol.[146]

212. [Alcohol]

From a mixture of pure and very strong *xjnf* with 3 *qbsut* of *tbmu*,[147] cooked in the vessels used for this

business, there comes a liquid, which when set on fire and while still flaming leaves the material [underneath] unburnt.

213. On leveling, or the measurement of heights

First, construct in this way an orthogonium [i.e. an instrument in the form of a right-angled triangle]: Make three flat straight rods, the first 3 [units—either] inches, feet, or cubits [in length]; the second, 4; the third, 5. Set up vertically the one that is 3 units long; lay horizontally the one that is 4 units long; and stretch the one that is 5 units long between the tip of the one that is vertical and the tip of the one that is horizontal. These three rods fitted together at the corners will then make the orthogonium. Now the vertical rod is called the perpendicular; the one laid flat, the base; the one stretching between them the hypotenuse. Then take a post whose height reaches up to the level of your eye; and attach the orthogonium to this at the middle of its base. Then place your eye at the corner where the base and hypotenuse join. Take a sight up to the corner where the hypotenuse and perpendicular meet; then walk forwards and backwards [with the instrument to your eye] until in your view the corner where the hypotenuse and perpendicular meet appears to join the top of the object whose height you are trying to find. When this has been done, measure the length of the distance from the place where you then were standing to the foot of the object. Subtract a quarter from this length. The height [of the object] is then obtained as the length of the remaining three parts plus that of the post that you were holding in your hand. Now you must watch carefully to see that the orthogonium with the post attached beneath does not incline in any direction: in order to detect any inclination, hang a plumb-bob down from the center of the hypotenuse. When it touches the center of the base, you will know that the orthogonium is not inclined.

214. The stone orchus or orebus

The stone *orchus*,[148] which the Alexandrians call cadmia, is engendered in humid places; and it is easy to pound up. Now it is black and it is an ingredient in silver solder.

---

[144] Chaps. 207 and 208 are near duplicates of 115 and 116.

[145] See chaps. 15, 18, and 83 and the note to chap. 15.

[146] An obol is 1/6 of a dram; approximately 1/6 pennyweight. Silver-tin alloys with 1/18 and 1/32 of tin melt respectively at 925 and 940°C, the higher melting point being appropriate for pure silver. They are whiter in color than comparable silver-copper alloys, but because of the formation of refractory tin oxide they would be less satisfactory as solders than the silver-copper-zinc alloys of later times.

[147] A simple code, known by cryptographers as the Caesar mono-alphabetic cypher, is used in the Latin. Its deciphering is simple: just substitute the preceding letter in the alphabet:

Latin:	xknk	qbsuf	tbmkt
	vini	parte	salis
English:	xjnf	qbsut	tbmu
	wine	parts	salt

The English version follows the original in failing to encode the *n* in the first word. This chapter provides one of the earliest

known accounts of the making of alcohol. The salt added to the wine before distillation would aid the separation by lowering the vapor pressure of water relative to that of alcohol. For a discussion of the history of alcohol, see Diels (1913, 1924), Degering (1917) and Forbes (1948). The translators wish to acknowledge a helpful correspondence on this chapter with Mr. J. J. McCusker of the University of Maryland.

[148] Svennung (1941: p. 83) has a discussion of the words *orchus*, *orebus*, and *erebus* in which he imaginatively suggests that *orchus* is a corruption of greek *orobus*, literally peas, combined with overtones of *orobitis*, gold solder, which suggests *chrysocolla*, the green mineral elsewhere used as a solder.

215.   The stone atriatis

The stone *atriatis* which is called *leocopandium* [white pandium]. There is a green earth in which it is engendered. When the earth grows and flowers again, it produces a white flower, convex, four-sided, with sharp edges. After this it shrinks and it will become a [crystalline] stone. But if the green earth contracts the flower when it is in full bloom, then [irregularly textured] rocks will result. Some of these are of a golden color, others of apple green, others *pandius*, others white. When these are struck, they give off fire. Quicksilver issues from them. In the month of April and May, when the earth is warming up and the flowers are abundant, excavate a humid place knee deep and the earth will be disclosed; you will find old flowers, hardened and sticking to the earth, which have become gemstones. Now others had flowered and hardened, but did not stick to the earth and have remained like pearls, because they did not encounter the right season. Others flowered in the appropriate season, like white snow. When you find these, lift up the barren earth with the flowers and put them into marble mortars, and when these are filled, put in water, stir well, and throw out the earth that is in the water, and there remains quicksilver. Some silver ore will also come out and runs when it begins to burn. Artisans collect it.[149]

216.   The pumice stone

Pumice stone is engendered everywhere: after it is ground, it is put into a new cooking pot and is placed in a potter's kiln, and is heated well, carefully covered so that no impurity can get into it. After this, it is taken out and ground for use, and it becomes part of the recipe for gold, substituting for a gem in the tempering of *calaina*.[150]

217.   The recipe for orpiment

1 part of clean ground orpiment, 1 oz. of quicksilver, 1 *tremis* [= ⅓ solidus] of gold. Beat the gold and make it into a leaf, and put the leaf and the quicksilver into an iron ladle, and heat it until the gold melts and mixes with the quicksilver. Afterwards put a little orpiment into the same ladle with the quicksilver mixture, and cook well, and shake it until it becomes *pandius*.

218.   Gold solder for fistulas

Gold solder for fistulas from gold strip: take one ounce of *asticum*, 1 oz. of copper calcine, 1 oz. of *afronitrum*, 1 oz. of olive-oil soap made without lime, 2 solidi of vitriol [*vitriolum*], half an ounce of vinegar, 1 oz. of water. Break them up and mix them, the copper separately and all the rest into one. Compound a little [of these together] as a solder to make, when heated, tubes from gold strip.[151]

219.   Gilding copper, silver, and brass

Encaustic[152] or the first burning-on [of gold amalgam] on silver, copper, and brass. Hammer out some gold, and make it into fine, thin leaves, and next put in some quicksilver, and melt the leaves until all the gold is melted. Now, if the quicksilver is diminished add more, until all the gold is cooked. Then put it in an earthenware pot and grind it with another piece of earthenware until the gold is attenuated and completely mixed with the quicksilver. Then scrape the vessel which you have to gild, and coat it with a little [of the amalgam], heat it, and press it out with a clean linen cloth, and wipe off all [the excess quicksilver]. Now put [the vessel with] what remains on the fire and test it in the same way. And if it is a new pot put on one or two gildings. However, if it is once lightly coated, then rub it with a hot iron [burnishing tool], and it takes on color. Then rub it with bread crumbs to clear up its color.

219-A.   [Gilding iron]

Iron is also gilded in the same way, but first it should be treated with alum.[153] Take some vitriol [*vitriolum*] and a little salt, and some very sharp vinegar, in a cup; and coat the iron that you have to gild with this. And here you have the first gilding.

219-B.   Chrysography with [gold] leaf [S]

Take the truest saffron, and scrape off the flower for a time. Then take an egg, open it and pour off what first comes out. Collect the white as it issues into the saffron and grind it properly,[154] coat whatever you wish with it, and lay the leaf on top.

219-C.   Again, the chrysography of scribes [S]

Take quicksilver, mix it with gold as in gilding,[155] grind it well and put it in a cup; and lay it on embers

---

[149] This chapter cloudily discusses the growth of mineral crystals in the earth, and reflects the belief that such growth was analogous to the growth of plants. It is difficult to identify the specific minerals involved. Though gemstones seem to be involved at first, the reference to burning and to quicksilver hints at cinnabar.

[150] *Calina* in **L** and **S**. Perhaps calamine is meant, elsewhere *cathmia*, which we translate cadmia.

[151] This is a duplicate of chap. 141.

[152] *Enencaus, encause* **P**; *Cencaus,* **S**; *cemcausis* and *cencausi,* **L**,—all corrupt latinizations of the Greek *enkausis*. It is an unusual word for the amalgam-gilding process but obviously compares it to burned-on wax encaustic.

[153] This etchant would be more effective if the *vitriolum* were blue vitriol. See notes, to chap. 141, 146-G and 146-H.

[154] Reading *utiliter* as **S**, **L**, and **K**. **P** reads *leviter*, lightly.

[155] Follows **K**, *sicut in deauratione*.

until the quicksilver is dried [out] and the gold remains; put this in a mortar; grind it well with an iron pestle, until it becomes powder. Take saffron, and incorporate it by grinding; if there is an ounce of gold there should be 2 solidi of saffron, which you put into water and let it cook down. In the same way put water from gum into the composition; grind it for use, and put it in a flask and hang it in the sun; and take it out of the sun whenever you want. Write what you want with the reed pen with which you [normally] write. Make a similar composition with silver and with copper.

220. How cooked sulphur is made

Cook lard and take 2 pounds of its oil and 3 pounds of sulphurous earth. Grind this earth and put it into a cooking-pot, boil it twice or thrice, and pour it out onto a brick.

221. A recipe for afronitrum

A second recipe for *afronitrum* which is also used for soldering gold and silver or copper [*S*]. 1 pound of Egyptian natron and 1 pound of axle-grease soap [made] without lime. Grind these properly and mix them together. Then put it in the sun, or in a hot place; it is useful in soldering gold. For silver because of the softness of the metal, a softer composition is made, i.e., two parts of soap and one of natron.

221-A. A recipe from Brindisi[156]

2 parts of copper, 1 part of lead, and 1 part of tin.

221-B. Another recipe from Brindisi

Two parts of copper, one part of lead, half of glass, and half of tin. Mix them together and melt. Cast them according to the size of the vessels. It is also used, with *afronitrum*, for the soldering of copper.

221-C. A recipe for cinnabar [L]

A recipe for true, clean cinnabar. Take 2 parts of quicksilver and 1 part of native sulphur, and 1 part of clean urine. Take a very clean strong flask that will endure heat without smoke. Put into the flask the sulphur, ground and mixed with the quicksilver, 2 ounces short of filling it; but if it is a larger flask, it should be short 3 ounces. Mix and shake. Get ready a smaller glassworker's furnace, which should amply hold the flask [K], leaving a place where the

flask may enter. Split reeds and with them light the furnace. Leave another window so that the flames may breathe out all round the flask [K]. The sign of [completion of] the cooking is this: when you see that the flask has less purplish smoke and is making a color like cinnabar, stop adding fuel, for the flask gives a crashing sound from the great heat. When the cinnabar is thoroughly cooked leave it to cool.[157]

221-D. A recipe for verdigris

Take copper strips, and scrape them down well, and hang them over vinegar. Scrape off and gather the stuff that collects on it.[158]

221-E. The recipe for lulax, i.e. indigo [S]

2 parts of verdigris, 4 oz. of clean vitriol [*vitriolum*], 2 oz. of Egyptian alum, 2 oz. woad leaves; pound them cleanly by themselves. Combine the verdigris, vitriol and alum, and take a half ounce of olive-oil soap without lime,[159] and mix the three ingredients in it. Now thoroughly mix them and pound them again with the soap. Then take the properly pounded woad leaves, mix them with the above-mentioned ingredients, and triturate thoroughly and allow to rest for a day.

222. This is the recipe for it

Mix 1 pound of clean defrothed urine with the above ingredients and grind for a time. Put it in an iron cooking-pot if you have one, otherwise in an earthenware one, and cook it until it is reduced to a third. Next take well-pounded cooked gypsum,[160] a half ounce of it. Pick up the concoction, mix the gypsum with it and triturate for a time, and put it in a pot. Place it in the sun and, when it has hardened break off a piece and put it to dry.

223. The recipe for ficarin

Take 1 pound of very clean lac and cook it down with 5 pounds of defrothed urine and cook it down cleanly and do not let it boil unduly. Then take clean bones of a crab, and burn them cleanly, and grind a sufficient amount. Mix them in the lac.

---

[156] Chaps. 221-A, and 221-B are in *S* and *K* only; 221-C is in *S*, *K* and *L*. In *P*, the first two ingredients of 221-A are given, followed immediately, without break or caption, by 221-D. The association of this copper alloy with the town Brindisi later gave rise to the word bronze. The composition specified in this chapter would be rather brittle. Brindisian speculum metal appears in chap. 89-A.

[157] The urine in the second sentence is included only in *K* and in the second version in *S* (fol. 213v.). The entire chapter is missing in *P*. A recipe rather similar to this appears in Theophilus, I.34, in which a crashing sound is attributed to the uniting of the mercury and sulphur.

[158] In *P* this paragraph follows immediately after the "two parts of copper and one part of lead" in the beginning of 221-A. Roosen-Runge (1967:2: p. 38) inferred from this that the copper strips were alloyed and that the recipe was to make the mixed white and green pigment called *granetum* by Le Begue—the *gravetum* of our chapters viii & x. The earlier readings do not support this.

[159] *P* reads without salt.

[160] I.e., plaster of Paris.

Take some fine wheat flour soaked in water, and well soak 1 oz. of it to the proper consistency, and unite them, i.e., the crab bones and the lac, by pounding them well, and mix them with the soaked wheat flour; put them in a pot, and dry in the sun. From this you may make *ficarin*.

224.   Coloring [glass for] mosaics green [S]

Take 5 pounds of a lump of clean glass and 2 oz. of lead-free copper filings, and put them in a new earthenware pot. Put fire underneath, and in the lower part of a glassworker's furnace cook them down for 7 days, and after this take it out and break it up into small pieces and melt it again [S]. It will be green colored.

224-A.   The ore of glass and its cooking

The ore [i.e., raw material] of glass. There is a sand that is engendered in various places; and this sand is also engendered in the mountains of Italy. There is also a blackish stone having the color of glass [obsidian?]. This is the testing of it: take some of this sand and put it in crucible [L] to use it. Light a fire with coals. Glass will run from below your hand [*sic*] but a useless kind.[161]

225.   The ore of lead and its cooking

Lead ore is a dusky earth which is engendered everywhere, more, however, in hot places; and the stone which is engendered in it is green, not whitish, and it is a heavy ore. The test of the ore is this: take it and put it in the fire; when it boils and is molten, it emits sparks. The grass which grows in this earth always withers under the heat of the ore. It is gathered as follows: because of the sweltering heat of the sun excavate the earth to a depth of 3 cubits— the earth is weak, and while the excavation is being done, it dries out. Then it is smelted in the furnace in the same way that iron is, though lead burns more.[162]

226.   Another smelting of lead, from the same ore

The same ore is not dried, but continually as it is being excavated it is put into an iron furnace with charcoal and a slow fire. The fire is not urged [with bellows] until nightfall, but during the night it is urged until the fourth hour of daylight. Then it is resmelted so that it becomes clean: it is put into the furnace again, and is smelted with charcoal made from pine or fir for 3 hours and it will be worked as it should be.

227.   Another recipe for glass

Take some of the same sand, and wash it off because of the dust [to] let it lose color; build a glassworker's furnace, and make 2 pairs of bellows. For the first working of the glass cook it down, as in the cooking of pitch. Afterwards take out that first glass because it is useless, grind it small [S] and cook it again in the furnace, as pitch is customarily recooked.[163]

228.   How skins should be dyed purple [L]

Take a skin that has been stripped of hair and properly washed, and for each skin take 5 pounds of nut-gall and 21 pounds [15 lbs., S, L] of water and put the skin in it and agitate for a day. After this wash it well, and dry it. Then take Asian alum and put it in hot water. After it has settled, pour off the water and put in warm water again, and agitate it. Put into this composition one or two skins [or as many as you wish (S, L)] and take them out and wash them once. Now each skin should have half a pound of vermilion.[164] This is the first dyeing of them.[165]

229.   Dyeing purple

Put defrothed urine in a cooking pot, and place it at the fire. Wrap some vermilion that has been ground in a mortar in a linen cloth of loose mesh, put it into the pot as it is heating, and agitate it until whatever can come out of the cloth does come out. Put the residue that remains back into the mortar and grind it, wrap it in the cloth, put it into the pot while

---

[161] This chapter is in *S* & *L* only, though in *P* its title, *de metallo vitri et coctione*, is prefixed to the text of chapter 224. In *L* it continues, without interruption or rubric, into our chapter 227.

[162] Note the reference to the smelting of iron in this and the following chapter. Was a blast furnace involved? The hint of a liquid product is reinforced by the direct reference to "running [i.e., molten] lead, iron and other metals" in the large scale operation on firestone (*lapis focarius*) described in chapter 135. Johannsen (1933), after reading these statements in the Hedfors version of the *Compositiones Variae*, saw in them some ninth-century evidence for a blast furnace producing cast iron. Such a development is generally believed not to have occurred in Europe until the fourteenth century, and there is not at present any earlier archaeological evidence for either furnace or massive product. However, many of the furnaces producing wrought iron and steel would occasionally over-carburize the metal, and they could probably have been operated in such a way as to make a molten product consistently had this been desired.

[163] This chapter in *S* bears the title *Probatio autem metalli*— the testing of ores. Bellows were not used in glass furnaces, and perhaps a metal-smelting furnace for reworking slags was intended, rather than the making of a frit and remelting it. In *S* and *L* this follows immediately after 224-A without the interruption by the chapters on lead and without heading.

[164] This vermilion (*vermiculum*) is clearly not the modern inorganic pigment of that name but is probably the organic dye kermes. See chapters 175 and 176. For a discussion of the dyeing of skins, see Edelstein and Boghetty (1965) and the excellent annotated edition of the *Plictho* of Giovanventura Rosetti (1548) published by these same authors with an English translation in 1969.

[165] In *L*, the purple (end of chapter 229), green (232), and apple-green (233) dyes are designated the second, third, and fourth dyeing respectively.

it is heating, and agitate it until no vermilion remains in the cloth.

Afterwards sew up the skins like a wineskin and take the broth, i.e., of the above-mentioned mixture, one and a half pounds [S] for each skin. Rub it well [into the skin], and let it remain the whole night in the mixture. Now in the morning again make enough of the mixture and, after pouring out the broth, wash the skins and dry them. Sheepskins are dyed in the same broth that was used for the former skins, i.e., in the same preparation in which goatskins have been dyed.

230. Dyeing a skin red

Let a skin lie in lime for 6 days and put it in salt and barley for 7 days. Then let it dry, and afterwards knead it; then cook vermilion in wine, and put the broth into the bladders for an hour, and let them dry.

230-A. Dyeing a skin green [S]

Stretch the skin on a rack and scrape it on both sides with a razor. Take some salt with flour and honey; mix them together, let them ferment and let the skin lie there for a night or two. Hang it in the sun, and knead it; dye it with copper, and knead it.

231. Dyeing a skin green

Take the dung of a dog, a dove, and a cock: dissolve it into a broth, and put de-haired skins into it. Process them there for 2 days, then take them out, wash them and let them dry. Then take Asian alum, and remember to do with these as we taught above should be done in the case of the purple dye. Then take well-pounded weld, cook it with urine, and when it is cooked let it cool. Sew up the skins into the form of a wineskin, as we said in the case of the purple dye, and put the mixture into these skins, and rub them well, inflating them a bit, so that it has air; and process them thoroughly until the preparation is absorbed. After this pour off the composition, wash the skins once, and again take 4 oz. of *lulax* to each skin and 6 pounds of defrothed urine and when the *lulax* is mixed with the urine, put it on the skins, just as formerly you put on the weld broth; mix well, until the wetness in the mixture is absorbed and used up. Then pour off what is left over of the weld broth and the *lulax*, let it dry, and dye a sheepskin in it, as we said before in the case of purple dye. It will be green.

232. Again, dyeing a skin green

Take de-haired skins, as we said above, and process them first in dung, then in alum, and when they are taken out of the mordant, sew them into bladders. Then take a half pound of *lulax* and mix in 10 pounds of defrothed urine put it into the bladders, and process them thoroughly, letting in a little air, as was said

above. Now do this continuously for 4 days; and after the 4 days pour the mixture into sheepskins, process them for 5 days, wash them and let them dry.

233. Dyeing skins apple-green

Process skins in an apple-green dye in the same way. Treat them with alum, as we said before, and when they have been washed after the alum, sew them up into bladders. Afterwards take well-pounded weld, cook it with well-defrothed urine, and, when it has cooled, put the broth into the bladders and process them as we said before for 5 or 6 days. After this pour it out and dye the sheepskins as we taught above, and after dyeing wash and dry them.

234. Apple-green purple

Process skins in purple [dye] as above. Put them into alum. Wash them off and dye them with apple-green. Then mix kermes and put the mixture into the skins that you have dyed; and process them as we taught above.

235. The first pandius dyeing

In the first *pandius* dyeing, process the skins in the same way as was said above and treat them with alum. After the alum treatment, wash off the mordant and dip them in vitriol [*vitriolum*]; after the dipping wash them well. Then compound vermilion, as we taught above [chap. 230]; and put some of the broth of this concoction into bladders and process them in the usual way; and when the confection has been poured out, dip the sheepskins, wash them and dry them.

236. The second pandius dyeing

Now in the second *pandius* dyeing, when the skins have been processed as above and dipped in vitriol [*vitriolum*] and washed, put some weld broth into bladders and process them for 4 days.

237. The third pandius dyeing

In the third *pandius* dyeing, when the skins have been processed as we said above, take some kermes broth and put it into the bladders; agitate and process as we said above.

238. [Pandius again]

Take 2 pounds of ground thin red-sea coral of a good color, 1 pound of murex lac, and 2 oz. of calcothar; grind and mix them all together, and cook them with urine; and when you want to dye, put some of the broth into the bladders, after processing in defrothed urine; process for 2 days. After this wash them well, and dry them out.

239. [Pandius *again*]

Take madder, pound it well and cook it in a cooking pot with urine.   Add a little alum, mix them together, and allow them to cool.   Then strain the broth and put it into the bladders made of skins that have already been processed, and agitate well and process them for a day, and wash and dry them. After this take 1 oz. of weld broth and 1 oz. of *lulax*, mix them together, and coat the surface of the skins with them.

240. *Dyeing bones, horns, and woods green*

First scrape whichever of these you want and put them into Asian alum: treat bones with alum for 12 days; horn, however, for 9 days; and wood for 4 days. Then cook well some weld, and, while it is simmering, put whichever of these materials you want into it; and when it has cooled, mix some *lulax*, and put them into it, and leave for 5 days.   Afterwards take them out and wash them.

241. *Dyeing the same materials blue-green*

When dyeing these things blue-green,[166] treat whichever of them you want with alum as we said above, and make some *lulax* and put them in it—if it is bone, for 10 days, but if horn, for 9 days, and if it is wood, for 3 days [*S*].

242. *Dyeing the same materials apple-green*

In dyeing these apple-green, treat the things that are to be dyed with alum, as we said above; and cook weld with defrothed urine; and when it boils, put them in it.

243. *A pigment [like] cinnabar*

Now if you want to make a pigment like cinnabar, take 2 parts of cooked sinopia, 1 part of *siricum*. Mix them together and temper with water.

243-A. *Parchment from ox-hide*

If you want to make parchment from oxhide, put the hide in lime and let it lie there for three days. Then stretch it on a rack and scrape it on both sides with a razor.   After it has been scraped, let it dry. Make whatever you want by cutting with a little carving tool,[167] and afterwards paint it with pigments.

243-B. *The recipe for white lead*

Take very sharp vinegar and pour it into a jar until it is half full.   Then hang thin lead sheets over the vinegar in such a way that they are not touched by the liquid.   Seal the mouth of the jar carefully,

but let it have a breathing hole.   Place it either in the sun or in any other hot place you like where it can stand undisturbed.   After 15 or 20 days to allow them to mix, collect the efflorescence that is formed on the sheets [of lead], wash it in water and dry it in the sun.

244. *How* cebellinum *is made*

Cebellinum will be made as follows: take wood of the turkey oak or *deirinum* and clean the bark off its branches and adze its surface smooth, and bury it in a muddy place for 20 years.   Afterwards take it out of the mud and leave it in the shade for a year to dry out and work out of it whatever you wish.

245. *Gilding iron*

If you wish to gild iron, take equal weights of calcothar and Asian alum and salt similarly, and tragacanth weighing as much as all three: and mix them all together with water and very sharp vinegar; put the mixture into a copper pot; and let it boil for one hour in daytime.   Afterwards, when the iron has been wiped clean and carefully polished with pumice stone, coat it with this concoction in the place where you want to gild it; and when you have left it for a short time in this concoction, wipe the iron, and it will have the color of copper.   Then take an onyx stone and rub the iron with it.   If it should lose its color by this rubbing, dye it a second time; but if it refuses to take the gilding, mix it [the gold amalgam] in equal amounts with the preparation and apply it as we said before.[168]

245-A. [*A dressing for (gilding) cloth*]

[Put] size[169] made from an oxhide on whatever kind of cloth you want to work; if it has to be polished, rub it with an onyx.

246. [*A recipe for a gold-colored transparent varnish*]

5 oz. of linseed oil, 2 oz. of *galbanum*, 1 oz. of turpentine, 1 oz. of Spanish pitch.   Melt these 3 substances (i.e., *galbanum*, turpentine, Spanish pitch) into one with a little linseed oil.   Now afterwards mix some linseed oil and ox glue with 2 oz. of oriental saffron, 3 oz. of frankincense, 2 oz. of myrrh [*S*], 2 oz. of mastic, 2 oz. of pine resin, 2 oz. of early-blooming poplar flower, and 2 oz. of varnish; strain into a

---

[166] *Venetum*: the color of the Venetian sea. For a pigment of this color, see chap. 283.
[167] Reading *scalpellatura* for *scappillatura*.   Chapters 243-A and 243-B are translated from *S* as they are not in *P*.
[168] This process roughens the iron by etching and coats it with copper by electrochemical replacement.   See note to chapter 146-G.
[169] *Bluta*. For identification of this as size, see Hedfors, pp. 24, 132, 135.   In *L* this chapter is entitled *De deaurationem pallii* and the application of gold leaf is clearly implied.   In *S*, as in *P*, this is a continuation of the previous chapter.

copper bowl.[170] When they seethe all together, mix with them 1 oz. of cherry-tree gum. When all these have been united, make them boil in 3 ounces of linseed oil by measure. After the cooking strain them through a linen cloth and mix the above-mentioned substances (i.e., *galbanum*, turpentine and Spanish pitch), and if later it is at all defective so that it cannot be dried, add as much mastic as you want, namely either an ounce or half an ounce, and it will be corrected.

246-A. [*Making gold leaf*][171]

How the leaf should be made. Mix together into an ingot [? *clavum*] 1 oz. of Byzantine gold, and 1 oz. of clean silver. Purge it by means of lead and afterwards cast it. Then mix and beat out a sheet, and when it has been beaten thin, cut it [into square pieces each] weighing 5 Byzantine *tremis* [= 1⅔ solidi]. If one is long or short, equal it out in breadth and in length with a hammer. Eight leaves should result from the [initial] 2 oz. after they have been made equal in size. Heat them on the hearth, beat them, holding them in iron pincers, and while you beat them, they should be spread from inside to the outside, so that they appear thinned in the middle. When they have increased by a half, cut them with a small knife three times by measure to give four equal pieces. Fold them edge to edge equally, extend them and cut them with shears [to give 32 pieces]. The pieces should be placed on each other edge to edge on the hairs [of a skin][172] pressed lightly by hand and put in oil. 64 leaves have now been made from the above 8. Then make a pouch[173] out of [sheet] copper and always beat them in it; and place more copper for the beating, one leaf above and one below. And as you beat with a flat-headed hammer, strike as many blows on one side as on the other. And when they have increased a half, cut them and lay one on top of another. Next put them in oil and always fold copper [between them] and fit them together and beat them long enough. Repeat this

until out of the [original] 8 leaves 1028 are made.[174] Trim them with the shears and wrap the trimmed pieces in a linen cloth, since they ought to be heated in the furnace where the gold leaf is put.[175]

The furnace itself should be 2 feet high from the ground and there should be placed on its wall [as a cover] a perforated tile with 3 holes on one side and 3 on the other and one in the center. And another tile should be put half a foot above the ground, with a hole in the center. And at ground level a hole should be made [through one of the walls] through which the wood is put, and [higher up] in front, [another hole] through which the gold is put. And you should clean the gold well [by heating it] with ashes of cow dung mixed equally with salt, burnt and ground like the ashes. In the first [heating] you should put in old ashes, in the second, new ashes, and in the third ashes similarly sieved.

247. *How transparent varnishes ought to be put over pigments* [*S*]

3 oz. of linseed oil, 3 oz. of turpentine, 2 oz. of *galbanum*, 2 oz. of larch [*larice, L*], 3 oz. of frankincense, 3 oz. of myrrh, 3 oz. of mastic, 1 oz. of varnish, 2 oz. of cherry-gum, 2 oz. of poplar flowers, 3 oz. of

---

[170] *mosana*, which we hesitantly assume to be an early reference to dinanderie.

[171] This chapter, which appears in *L* between our chapters 233 and 234, is lacking in both *P* and *S*, but we include it for its technical interest. The Latin is unusually obscure and other interpretations could be easily justified. Theophilus (I.23) however, gives an excellent description of twelfth-century technique of making and applying gold leaf. For the history of sheet metal in general see Theobald (1912). The traditional technique of gold beating at its height, before the introduction of modern methods, is described in detail by Lewis (1763) and Ure (1842).

[172] *Capilatoras*, literally hairy. This is not the fine skin in which to beat the leaves as in later practice but is perhaps a short-haired pad to facilitate handling the leaves. See, however, Hedfors (p. 128) for a different interpretation.

[173] The use of copper plates to interleave the gold instead of skin (which Theophilus describes and has been standard for centuries) is mentioned by Pliny and even as late as the sixteenth century (Porta, 1589).

[174] *Sic.* Actually there should be 2^10 or 1024 leaves. Assuming each leaf to be 5 cm. square, the area per ounce of gold leaf would be 25,600 sq. cm., and the thickness 0.006 mm., or twice this for the 2 ounces if no silver had been removed. This is some fifty times the thickness of modern leaf (about 1/250,000 inch, 0.1 micron, equivalent in round numbers to 100 sq. ft. per troy ounce).

[175] Hedfors (1932) translates this sentence thus: "Schneide sie mit der Schere zurecht, alle Abfälle aber wickle in ein leinenes Tuch, damit du sie im nämlichen Ofen schmelzen kannst, der [das Gold] für das Blattgold aufnimmt." It is certainly plausible that a scrap-recovery process on the trimmings should be referred to, but there is no mention of melting in the Latin and we believe that the furnace is actually a cementation furnace for use on the metal in process, not the scrap. Melting was usually done in a hearth using bellows, while the furnace described, which is wood fired, is similar in both size and construction to the low-temperature cementation furnace referred to in Theophilus (Book III, chap. 33). Moreover, the leaves were of an alloy containing 50 per cent silver and would be white unless chemically treated to deplete the surface in silver and so restore a golden color. Reaction with the prescribed mixture of ashes and salt would do this effectively in an hour or two at temperatures far below the melting point, especially if heating were repeated three times with the specified ever-stronger mixtures. (See chap. 3, fn., for some similar processes.) The final leaves would be almost impossible to handle and would be too fragile to survive the operation intact, so it is likely that the cementation was combined with the first three annealing operations on the sheet while it was thicker. Some centuries later "gold" leaf for spun gold was made by cladding silver sheet with gold and hammering it down (Biringuccio, 1540, in Book III chap. 9). This would be more economical of gold but the product would be similar in nature and appearance to that made by superficially cementing the gold-silver alloy as suggested here. In *L*, this chapter continues into a paragraph describing the dyeing of tin leaves to represent gold (see chapters 115 and 116) and their application to glass, wood, and skin as well as to lead and tin objects.

almond-gum, 2 oz. of pine resin. All these should be pounded and sifted, then [put] into a brass dish with the above-mentioned linseed oil; and they should be put into a hot oven where they may be cooked without flame in such a way that they do not evaporate. And when they have been cooked, they should be strained through a clean linen cloth: and if they come out thin, cook them down until they become thick. Afterwards you will be able to varnish whatever kind of work you want, either paintings or carved work. When they are varnished, leave them to dry.

248. Chrysography

File some refined gold with a fine file, and put it in a porphyry mortar with very sharp vinegar, and grind it uniformly and wash it as long as it is black and decant. Finally put in either a grain of salt or some natron, and in this way it is made liquid. Afterwards write with it and polish the letters. All metals are made liquid in this way.

249. Writing in gold letters

Take gold sheets or silver ones and make them liquid by grinding in a mortar with Greek salt or natron, until they are impalpable. Then add water and decant [S], add salt a second time, and wash it in the same way. When pure gold remains, add a little verdigris and ox-gall, grind them together and write [with the mixture]. Polish the letters.[176] Now if you wish the work to be spread out and want to write more lavishly, grind separately 4 parts of cleavable orpiment and 1 part of *elidrium*, and when you have sifted it mix it with an equal amount of gold, grind it uniformly and write. When it has dried, polish the letters. You can also paint both on glass and on marble with this, in the same way that you write with [pure] gold.

250. Gilding a skin

Take a red skin and rub it thoroughly with pumice stone; then wash it with warm water until the water comes off clear; afterwards stretch it on a rack and scrape it up to four times; then stretch it on a place[177] that has a clean surface, and even it out thoroughly with a clean piece of wood. Now, after it has dried, take the white of an egg, dip a clean sponge into the liquid, and go over it once in stripes. Now if this is not enough, go over it again; and when it has dried, place a leaf [of gold on it], then dip the sponge in water and press the leaf to the skin; when it has dried, polish it. Then rub the top of it with a clean skin and polish it a second time. Gilding is done in the

same way with tragacanth [in place of egg white], but you should put it in water overnight in order to dissolve it.

251. The recipe for malta

8 pounds of olive oil, 8 pounds of cheese, 30 egg yolks, an oz. of egg white, half a peck of clean lime,[178] a pound of clean flax cut up very fine.

252. White copper

Put white copper into the bottom of a little cup, put glass on top, and so melt them together. When it is melted and you want to cast it, remove the glass with a little rod: it will not lose its color.[179]

253. Lac, how it is worked for painting on wood or on a wall

First grind lac and from it pick out the knobbly bits and impurities; then put it in a mill and grind it fine; then take the urine of a man or woman and first put it in a cauldron and let it boil until it is all reduced to a third. Continue always to take off the froth. Afterwards put in the lac and let it boil; then take very clean alum, and grind it and mix it in the above-mentioned lac. Then take a small cloth and keep on dipping it until a good color appears [on it]. Then put the liquid into little pots and work [with it]. Throw out the stone[180] which forms in the liquid, because it is of no value. Into 5 pounds of lac put 5 oz. of alum and 10 pints of urine.

254. Lime and sand

Furthermore a builder must know what are the practical qualities of lime and sand. Now, there are three kinds of sand that can be dug up from the ground: black, reddish, and grayish-white. Of all these the principal ones, of greater value, are the reddish ones. The grayish-white is second in value, and the black occupies third place. Of these, then, that [sand] which emits a strident sound when squeezed in the hand will be useful for the builder. It is excellent also if, when it is spread onto a cloth or the linen part of a white garment and shaken out, it leaves no stain or dirt. But if there is no sand of the dug type, sand will be collected either from the gravel [S] or the banks of rivers. Sea sand dries out more slowly, and therefore should not be built with

---

[176] This is the title of the chapter in **S**.

[177] Reading *aream* as in **P** instead of *axem* as in Phillipps's transcript.

[178] **S** reads *ulcis mundium dimidium*, which is meaningless and **P** changes to *calcis mundae modium dimidium*. The material seems to be a kind of cheese-egg gesso or sizing material. Malta is probably a corruption of *multa*, mortar, as in chapter 103.

[179] The cover of molten glass is used to prevent oxidation or evaporation of arsenic from the alloy and consequent loss of whiteness.

[180] Probably a crystalline lump of superflous alum or uric acid compounds.

continuously, but intermittently, so that its weight does not wreck the structure. It also loosens the plaster on vaults because of its deceptive wetness. For the sands that are dug up are particularly useful for plaster and for vaults because of their quick drying. They are better if they are mixed as soon as they are dug out, for they deteriorate in sunlight, or in winter, or in rain. River sands will be more suitable for plaster. But if it is necessary to use sea sand it will be advantageous to immerse it first in a pond of fresh water so that it may lose the defect of saltiness by being washed off in fresh water.

Now we make lime by calcining hard white [lime-] stone or travertine or dove-colored[181] [lime]stone from a river also, or red stone, or a spongy one [S], or, finally, marble. Whatever [lime] comes from dense or hard stone is suitable for buildings, but that from porous or softer stone is more usefully applicable to plaster. Now, in two parts of sand, one of lime should be mixed. But in using river sand, if you add a third part of crushed chalky brick, the solidity of your structures will be wonderfully pre-eminent.

255. Brick walls

But if you want to make brick walls in a very large house, you should take care that, when the walls are topped off, a terracotta structure should be made that is to be beneath the beams, with cornices a foot and a half high so that the rain will not penetrate the wall if the tiles should be broken or become rain-sodden. Then you must take care that the plaster is laid on the brick walls when they are dry and have been roughened, because it will not adhere to wet or smooth ones. Therefore you first have to scratch lines on them three times [in three different directions] so that they hold the plaster without its breaking off.

256. The recipe for sapphire [-colored glass]

For staining [glass] sapphire, silver and sulphur should be set on fire together; afterwards from 2 pounds of clear mosaic glass and 3 oz. of the above-mentioned sapphire cooked together, you will make jacinth stones.[182]

257. The recipe for red glass

Take iron clay and cook it on the fire, and afterwards pour wine on it. In order to produce a red

vein, crush it with a pestle after mixing in some copper; and so melt with glass [S] in a furnace.

258. Another way

Pour crushed Ardennes stone over green glass with which copper has been mixed.

259. Silver solder and gold solder

Take 2 parts of silver and a third part of copper, add a little tin and you can join silver or copper well.

260. Another one

Also, take three parts of gold, a fourth of copper, and melt them together.

261. Another tin solder

Mix together 2 parts of axle-grease, with a third of resin and an equal amount of tin filings. You will be able to solder [with the mixture] if you heat it gently at the fire.

262. [Gilding with] a pennyweight of gold

A plate of copper which is 10 inches wide and equally long can be gilded with a pennyweight of gold.[183]

263. Quicksilver

If quicksilver has stuck to a gold object that you cannot put on the fire [to remove the mercury], take the urine of a man, mix with it vitriol and salt, make a thick paste, put some of it on any of the quicksilver that has stuck to the gold object, and leave it on for some time. Then wipe it off and the quicksilver will not be visible. Then rub it with a [burnishing] tool, as you know how.

263-A. [Silver solder]

Weigh out 2 pennyweight and an obol of the purest copper, and 1 pennyweight of silver, and melt them together. Next, hammer it to the thinnest that you can; and afterwards thoroughly burn some winestone; and when you want to solder, take some of it and temper it with water but leaving it thick. Then cut up very finely that very thin copper and put it in and on top of the joint that you want to solder. Then add on top the thick paste that you made from the water and wine-stone, put it on the fire and work the bellows.

263-B. [To color gilded work]

Take a lot of vitriol and ignite it well, and some good salt, so that there are two parts of salt and a

---

[181] Cf. Biringuccio, 1540: p. 148: "la prima di tutte l'albazano, [l'altra], el trevertino, la terza e la pietra columbina biancha come marmo. . . ."

[182] The use of silver to color glass is described in detail by R. H. Brill (1970). It is not necessary to melt the materials, since diffusion of silver from the surface into the solid glass occurs readily at low temperatures and produces deep color.

[183] S reads "with ten pennyweights of gold."

third part of vitriol. Afterwards mix them together and temper them with the best vinegar. Then wash the gilded work well and coat it all over with the above-mentioned porridge. Next put it on the fire and heat it until it becomes red hot and then quench it in a copper pot and wipe it off with a bristle brush. Now, if it [the poor color] does not disappear the first time, apply salt, and repeat.

264. The lead arrow, for setting on fire

Melt lead once, twice, or three times, and clean it of all dross, and leave it until it is collected v.n.n.[184] After this bring up a tray, properly pound the lead in it, soak it with vinegar, then take off the scum that it gives out and dip the arrow in it. Sharpen an arrow with this scum on [a plate of] lead, as on a whetstone, until it is shining bright; and the arrow itself will now be thoroughly coated with lead.

265. Another toxic substance, with which an arrow may be poisoned in battle

Take the sweat which appears between the hips of a horse on the right side, and dip the arrow in it. This has been properly proved.

266. The arrow which emits fire

The arrow for emitting fire is triple-spiked and perforated. Now the recipe for the fire is as follows: 1 oz. of naphtha, 2 solidi of tow, 4 solidi of seasoned pitch, 1½ solidi of a clean drop of native sulphur, 1½ solidi of *climatis* [S], 1 solidus of sea salt, 1 solidus of olive oil, 1 solidus of raw bird lime, 1 solidus of jet stone, 1 solidus of soap made from olive oil. Take these, put them in a marble mortar, and grind them properly with an iron pestle.[185] First finely grind the naphtha that you have put in with the tow, the pitch, the *climatis*, and the native sulphur, the jet, and the sea salt. After this put in the olive oil, soap, and 1 solidus of a woman's milk, and unite everything together by grinding finely. Thoroughly grind four nut kernels and make your composition. Now, the milk should be fatty.

This is the way of dipping [i.e., coating] the arrow: dip some soft flax tow [in the incendiary mixture] and make a fine little rope such as can fit into the

perforations. With the remainder coat the arrow as you have provided for; and when you stretch your bow set the arrow on fire and immediately shoot it where you want a fire to be started by the arrow.

267. Another [recipe], a short one

Another kind of arrow, which emits fire with a simple recipe: native sulphur, colophony, and resin in equal weights, and nut oil. Mix these together and dispose it as we said above.

268. Another kind

The ballista with the arrow [projectile] envenomed in toxic. Line the channel with copper so that it does not catch fire. Now on the projectile, treat the triple-spiked head in accordance with the former recipe. There is also an incendiary stone which emits fire: this is not smooth but rough and has perforations to contain the composition.

269. The fastest [incendiary] composition

1 pound of sulphur, 1 pound of naphtha, 1 pound of tow, 1 pound of *climatis*, 1 pound of seasoned pitch, 1 oz. of a woman's milk, 1 oz. of hog oil, 1 oz. of jet stone [1 oz. of resin (S)], 1 pound of seasoned resin, 1 pound of native sulphur, 1½ oz. of liquid pitch, 1 oz. of cedar [pitch], 1 oz. of olive oil, 1 oz. of cooked sulphur, 3 oz. of orpiment, ½ oz. of natron: collect all the dry ingredients and grind them for a time, then take all the moist ones and mix them all together; then grind carefully for use, and when they have been compounded use it by coating the stone with it, filling all its perforations. Put it into the ballista, apply fire to it and shoot it off quickly.

270. The construction of a battering ram for taking walls by assault

Make three front feet 5 cubits long, middle ones 4 cubits, rear ones 3 cubits. And there should be wheels one and a half spans high, and 4 inches thick. Make them round and make a hole in the center, cut columns, and insert them 4 inches deep into the wheels. [Cover the wheel (S)] and make a joint on top and fix it tightly with a wedge [? *meura*]. Bind the rams and wind ropes around them. Shield with leather, and cover on top with pieces of felt, and over the felt, put pieces of leather; and over the leather 4 inches of sand, and over the sand, wool, so that the sand cannot move, and on top [more] pieces of leather. The columns themselves should have hinges such that when it walks you may turn it where you wish. If you tighten the hinges it cannot move, because they are joined together inside. Now put supports underneath for the wheels. You can join

---

[184] The meaning of this abbreviation is unknown. Chapters 264 and 265 have been cut out of S and are not available for comparison.

[185] In S, fol. 41v, the balance of this chapter has been cut out. For a discussion of the history of military incendiarism, see J. R. Partington (1960). Partington believes that the famous Greek fire was a volatile distilled oil projected by a flame thrower, quite different from the sticky napalm-like mixtures of these recipes, which are all excessively complicated and clearly represent many stages of corruption of words from several languages unknown to the scribes.

up to a wall with this engine, and you can work without hesitation.[186]

271. How you should set the shield of a battering ram on fire

Fill an unfired pot with the composition *dedamia*,[187] light it and toss it on to the shield and so set fire to the leather pieces and the wool. After this there remains the sand fixed in place because it does not catch fire. Then throw stones to make the sand fall down in heaps; again make a similar pot of the same composition [and] throw it on the shield; and if, because there is too much sand, it [still] does not catch fire, throw more stones at it and fill another pot with the composition and throw it on.

272. Four types of [incendiary] material

The most useful incendiary composition contains four types of material: naphtha, pitch, tow, *climatis*.

273. The recipe for naphtha is this

Naphtha [is made thus:] namely, the stuff washed out in processing flax or olives mixed with [I][188] 1 pound of the cleanest naphtha, [II] 2 pounds of native sulphur, 4 oz. of *milinum*, [III] 4 solidi by weight of sea salt, [IV] 4 solidi by weight of 2-year-old oil, from freshly collected and pressed olives (*S*), [V] 2 oz. 1 solidus of clean, liquified colophony, [VI] 4 solidi and 6 *aurei* of pitch oil, [VII] 1 oz. of hard pitch, [VIII] 1 oz. of serpentine turpentine oil, [IX] 2 solidi of cedar resin, [X] 2 solidi of cypress pitch, [XI] 1 solidus of mastic, [XII] 1 solidus of jet stone. These are the twelve ingredients in the recipe for naphtha.

274. The recipe for serpentine oil[189]

1 part of turpentine oil, 1 part of ordinary oil, 1 part and 2 solidi of oil of laurel.

275. The preparation of these [ingredients] is as follows

Grind all the dry ingredients, afterwards thoroughly mix the wet ones with the powder of the dry ones, and rub them all well. After this put it in a glazed

earthenware pot and leave it like this for 2 or 3 days, and after this, take it out, heat it a little until it boils, and leave it to settle.

276. The recipe for pitch

Take [I][190] 1 oz. 1 pennyweight of dry pitch, [II] 1 pennyweight of native sulphur, [III] 1 pennyweight of resin, [IV] *iscira*, [a] [V]. Then put in 1 oz of *amor aquae* 1 oz. 6 *aurei* of clean balsam, [VI] 1 solidus of *silicum* oil—which others call *vicinum* [castor oil], others *lancidum* [oil of asphodel] and others *viscum* [oil of mistletoe]—[VII] 1 solidus and 12 *aurei* of cypress pitch, [VIII] 1½ solidi of pine pitch, [IX] 2 solidi of olive oil soap, [X] 1 solidus of natron, [XI] 1 oz. of dried and ground parsley [*selinistreum*],[b] [XII] lavender [*alochias*], 4 solidi, 1 pennyweight of it dried and ground,[c] [XIII] *convulvus* [*robasticis*],[d] take one solidus of its seeds, dried and ground.

[Keep] all the dried, ground ones apart; then mix them together and add to them 2 solidi of hornbeam and mix everything and unite by grinding. Put it in a glazed earthenware pot; and when it once boils over a slow fire, it will become a compound like pitch.

[a] Others call this *flos aquae* [flower of water), others *oleum aquae* [oil of water], others *celidonia* [*cicedonia*, *S*], but the Alexandrians, *amor aquae*. And it forms in water where there is white, reddish, or black earth. And when it comes out, the water will become a yellow-green on top of the water around where the waters flow away. Now, it is a heavy sickly liquid owing to the earth that generates it as an efflorescence. And it is collected as follows: it is collected, if the place is hot, from the months of March or April until October. Take a very soft washed wool and lay it on top of the water, and squeeze it into a glass pot which has a very tiny hole in it as if made with a needle; seal it with wax and set it undisturbed in the sun for ten days and nights. After this open up the hole by removing the wax, holding clean wool beneath it until the water soaks into it, and *amor aquae* remains.

[b] Others call this *rodica*, others *pancium* [*pantium*, *S*, *pandium*, *L*] others *rusticum*, others *gumma* [gum], others *manticam* [*S*, *L*], others thyme, others *trichas*, others *tricoselinon* [maidenhair] for it grows in water like parsley and on walls where there is lime and it is called the hairy plant.

[c] Others call this *sticis*, others *calmidam*, others *cathan* [*caddian*, *S*], others *ageropa*, others *marcianin*, the Alexandrians however call it *scaramandia*, the Isaurians [call it] *papati*. It is a whitish slender plant with branches and involute spines. Many people light candles out of it. It grows in rough, rocky places. Its leaves are thick and like those of myrtle.

[d] Others call this *exmilax*, others *ptelacion*, others *quisnasbatu* [*quinosbatu*, *S*. *Kunobatos*, Dioscorides],

---

[186] This whole paragraph is more than usually obscure.

[187] On the nature of *dedamia*, see chap. 162. In *S* (fol. 42v. lines 15-18) there follows a paragraph on *dedamia* that is utterly beyond our powers of interpretation. It reads: *Compositio dedamias electei pugnatoris. Compositio antimilli & ido tuthora cum macu bellero fantiu per seunos cyronon elandi. onos. milionos. Incompositione alentiu gentio. nossantonio. & abrochiu. aquilino.*

[188] The interpolated Roman numerals are all from *S*.

[189] In *P* this definition of *dracontoides* (which is not in *S*) occurs at the end of item VIII in the middle of the above recipe and bears a separate chapter heading. We transpose it to maintain the main sequence of items I–XII, as in *S*, but retain Phillipps's heading and chapter number.

[190] The Roman numerals that identify the various stages of the composition are in *S* only. There is considerable variation of spelling in the different manuscripts and most of the names are so corrupted as to be unrecognizable. The material in notes *a*, *b*, *c*, *d* are interpolations in the text, here separated for clarity, as also in chapter 277. The symbol ꝺ, a crossed Greek δ, denoting denarius, appears in *P* for the first time here. *S* has simply a crossed *d*. The substance *amor aquae* is clearly petroleum collected from natural seepage. (See footnote *a*).

others *cucudera*. It grows like a bramble, but with stronger branches and is dense; and its fruit is like the jujube but more rounded, for which reason countryfolk call it the wild jujube; and it has hairy, triangular seeds inside.

277. *The recipe for tow*

1 oz. 3 pennyweight of milk of iron, 1 oz. 12 *aurei* of native sulphur, 3 pennyweight of Persian alum, 1 solidus of maple gum, 1 pound of *amor aquae*, 4 oz. of balsam, 1 pennyweight of olive oil, 4½ *aurei* of jet stone, 2 *aurei* and 3 grains of hornbeam, 7 solidi 2 *aurei* of fat cedar resin, 4 oz. of soft [*S*] apple-green sulphur, 12 *aurei* of oil of laurel, 4 *aurei* of clean apple-green turpentine resin, 1 oz. of cold pine pitch. *Apallis.*ᵃ This is a tall plant reaching up to the knee, with leaves like those of myrtle but much larger. There is another plant, one is the greater *titior*, and another slender one the lesser *titior*, a span high; and when you break the greater *titior*, milk will come out. Its branches are rounded and its leaves thick. Collect and dry the milk [take] 4½ *aurei* of it. Milkweed [*Bracchia*].ᵇ Collect, dry, and grind 1 oz. and 1½ *aurei* of this. Seafoam [*Onia*].ᶜ Take the skin and dust of seven of these fungi. Take 1 solidus of dried, black, ground hellebore, 1 oz. of the gum from the tree *elaton*, which is the fir tree. Grind and mix all the dry ingredients separately and the wet ones separately, then mix them together, put them in a glazed earthenware pot and let them boil over a slow fire.

ᵃ Others [call this] *ramitan* [*yaminthan*, *S*], others *cordenan*, others *daucallis*, others *maragnin*, the Egyptians call it *fondella*, country folk call it *tinctio* [dye], others *polligalla*.
ᵇ Others [call this] *tutumallum*, others *daucia* [*S*], others *leptugalia*, others *polligala*, others *leptotitiu*. Now it is like the former, but not so tall, for the former throws out branches as soon as it begins [to grow], while this one has rounded branches above, slender leaves, but more rounded than the former, which is why the Ethiopians called it *surgana*.
ᶜ Others [call this] *laucia*, others *sehum marinum* [sea tallow], others *briania*, others *spuma marina* [seafoam], others *eleoboron*, others *magentia*, the Egyptians *drantia* [*P. daucia*, *S*]. It grows everywhere, more however in very hard ground. It is a rounded fungus. Country folk call it amanita. When it is dried and you hit it, it gives rise to lots of dust, so they call it *girogua* [*S*]. It grows in the earth completely round, and if, when it is dry, it is stepped on or hit, a dust will come out and the residual hard skin remains like a cooked egg.

278. *The recipe for* climatis¹⁹¹

4 solidi and 3 *aurei* of dried and ground flowers from the twigs of a wild vine, 1 oz. of milk of iron, 1 oz. of naphtha, 1½ oz. of native sulphur, 2 solidi of resin, 2 solidi of olimpian pease. Mix all these together and put them in a glazed earthenware pot, and boil them once over a slow fire and allow to settle.

¹⁹¹ Title in *S* reads *Antimis brionia clymatis*.

278-A. *The more useful incendiary mixtures of the four above-mentioned materials, i.e., naphtha, tow, pitch, and* climatis

4 pounds of good wet naphtha, [3 pounds of cooked down wet pitch, 2½ pounds of cooked wet tow, 2 pounds of cooked wet *climatis* (*S*)]. Mix them all in a copper cooking pot, and cook over a slow fire, and fill it up [?] from the copper pot. This is the foundation of incendiary compositions [*S*].

279. *An aid for extinguishing* [fires]

If a fire blazes up, it should be extinguished with sand and bran [*S*], if it blazes up further, put on sand soaked in urine.

280. *How soap is made from olive oil or tallow*

Spread well burnt ashes from good logs over woven wickerwork made of tiny withies, or on a thin-meshed strong sieve, and gently pour hot water over them so that it goes through drop by drop. Collect the lye in a clean pot underneath and strain it two or three times through the same ashes, so that the lye becomes strong and colored. This is the first lye of the soapmaker. After it has clarified well let it cook, and when it has boiled for a long time and has begun to thicken, add enough oil and stir very well. Now, if you want to make the lye with lime, put a little good lime in it, but if you want it to be without lime, let the above-mentioned lye boil by itself until it is cooked down and reduced to thickness. Afterwards, allow to cool in a suitable place whatever has remained there of the lye or the watery stuff. This clarification is called the second lye of the soapmaker. Afterwards, work [the soap] with a little spade for 2, 3 or 4 days, so that it coagulates well and is dewatered, and lay it aside for use. If you want to make [your soap] out of tallow the process will be the same, though instead of oil put in well-beaten beef tallow and add a little wheat flour according to your judgment, and let them cook to thickness, as was said above. Now put some salt in the second lye that I mentioned and cook it until it dries out, and this will be the *afronitrum* for soldering.¹⁹²

281. [*A white glaze*]

The white pigment: 10 pounds of tin and 1 pound of lead reduced to powder [by calcination], 10 pounds of Asian alum, 8½ pounds of yellow sand. Make a furnace and fire it, and after the cooking break up the cooked stuff and sieve it. Next add to it 9½ pounds of similarly sifted sand, and then 6½ pounds

¹⁹² *Cf.* chap. 141.

of lead and 5 pounds of tin, [calcined] in the same way as described.[193]

282. *Starch paste*

Starch paste is half a pound of wheat kernel with mixed hot water, and 5½ ounces of glass sapphire[194] and sufficient water.

283. *The blue-green pigment*

The blue-green pigment: 10 oz. of cleavable alum and 5 oz. of sand and 4 ounces of white rounded stones, all of them roasted and sifted, and 4 ounces of lead, and 1½ oz. [of ____?], and 15 [oz.] of silvery [– –?].

284. *The stone from* Egrippus

If you have found the stone, from the earth of *Egrippus*, after it has been roasted and sifted, add to it 4 pounds of lead, and mix into this composition 1 pound of purple earth. Also, add 2 pounds of rock from Corinth and 1 of lead and 1 pound of purple earth.

285. *The recipe for sesame candy*

The recipe for sesame candy. Put white pure honey near a moderate fire in a tinned [pan] and stir it unceasingly with a spatula. Place it alternately near the fire and away from the fire, and while it is being stirred more extensively, repeatedly put it near and away from the fire, stirring it without interruption until it becomes thick and viscous. When it is sufficiently thickened, pour it out on a [slab of] marble and let it cool for a little. Afterwards, hang it on an iron bolt and pull it out very thinly and fold it back, doing this frequently until it turns white as it should. Then twist and shape on the marble, gather it up and serve it properly.

286. *Sugar candy*

Now by a similar cooking process [put] some sugar soaked in a little water in a tinned [pan] and defroth it when it boils and strain it well in a colander. In this way, after adding in the ingredients that you know, stir it unceasingly until it reaches [the correct] consistency. Pour it out in separate pieces on a marble [slab] that has been lightly oiled. Carefully cool the pieces on the marble, separate them from it by hand and keep them properly.[195]

287. Penidias *candy*

Now *penidias* candy [is made] like sesame candy after the sugar has been defrothed and strained, but without stirring it. When it has been fully cooked, work it on the bolt as described above, then shape it by cutting with shears.

288. *How azure is milled*

Mill azure with soap; afterwards wash it well in water.[196]

288-A. [*On weights*]

Hence it was agreed that a *sexta* should be made and called a *siliqua*. The smallest is the *calcus*, weighing the same as two seeds of the chick-pea.[197]

---

[193] This is the white marzacotta glaze later so popular in majolica and Delft ware. Chaps. 283 and 284 below seem to be garbled versions of colored lead glazes for use on pottery, though they could perhaps be ground and used as pigments.

[194] *Vitreum safirum. Cf.* chap. 256, though the sapphire glass there described would hardly be a useful ingredient in a paste.

[195] On the history of sugar, see M. F. Deerr (1949). Although sugar was known in classical antiquity it was not much used in Europe until the twelfth century, after Arabic influence.

[196] This seems to be a reference to the flotation process for extracting pure ultramarine blue from ground lapis lazuli by mixing it with a greasy compound and kneading under water. It depends upon the fact that the interfacial energies cause grease to stick preferentially to sandy and earthy particles, leaving the ultramarine, which is preferentially wetted by water, free to float in suspension. A pure soap alone would not work, but some unsaponified fat in it could make it effective. Later recipes call for mixtures of pine resin, linseed oil and mastic. The earliest clear description and one of the best is that in chapter 62 of Cennino Cennini, *Il Libro dell'Arte* (*ca.* 1400, English translation by D. V. Thompson, Jr. [1933]). The anonymous Bolognese manuscript *Segreti per colore* of slightly later date published by Merrifield (1849: **2**: pp. 340–382) has even more circumstantial detail.

[197] These two lines are hexameter and seem to be part of a longer mnemonic poem on weights.

288–B. [A table of runes][198]

ᚨ	a	ag	[= *ac*, oak]	ᚿ	o	os	[= *os*, mouth]
ᛒ	b	berch	[= *beorc*, birch]	ᚠ	p	perd	
ᚻ	c	cen	[= *cen*, torch]	⯑	q	cui	
ᛘ	d	derhu	[= *daez*, day]	ᚱ	r	rat	[= *rad*, wagon]
ᛗ	e	eg	[= *eoh*, horse]	ᛋ	s	sigil	[= *syzil*, sun]
ᛈ	f	feu	[= *feoh*, cattle, property]	ᛏ	t	tir	[= *tir*, Tyr]
ᛉ	g	genue	[= *geofu*, gift]	ᚢ	u	ur	[= *ur*, bison]
ᚼ	h	he	[= *haezl*, hail]	ᛡ	x	xen	
ᛁ	i	—	[for *is*, ice?]	ᛢ	y	uir	[= *yr*, bow]
ᛦ	k	cer					
ᛚ	l	lag	[= *lagu*, lake, sea]				
ᛗ	m	man	[= *man*, man]	ᛘ		[z]	
ᛏ	n	net	[= *nied*, necessity, oppression]	ᛜ		[ng]	
				ᚦ		[th]	

In addition there are:

---
[198] See Appendix C.

288–C. [A table of weights][199]

ϒ	Twelve oz., 12.	Half an obol,	𝕀𝕀
ᛋᛋᚵ	Eleven oz., 11.	One obol,	⏑
ᚵᛋᛋ	Ten oz., 10.	Two obols,	⏑.
ᚵᛋᚵ	Nine oz., 9.	Three obols,	τ.
ᚵᛋ	Two-thirds [pound =] 8 [oz.]	Four obols,	f.
ᚵ	Seven oz.	Five obols,	6.
ᚵ	Half [pound = six oz.]	6 obols represent a pennyweight	
ᛜᛋᚵ	Five oz.	Z[200] 8 siliquae,	H.
ᛜᛋ	One-third [pound = four oz.]	One solidus,	Ħ.
ᚵ	One-quarter [pound = three oz.]	Dram, i.e. ounce	
ᛋ	One-sixth [pound = two oz.]	Half a solidus,	℔.
⏗	One ounce.		

288–D. [Making French soap]

Agitate with cold water two parts of oak ashes with a third of oak [*sic*] lime. Afterwards when they are well stuck together, put the whole in a basket, strongly pressed down to make on top a place for the water so that it does not run away. In this you will put cold water two or three times according to the amount consumed by the underlying ashes and the lime. Not quickly but on the following day, the water will drip down onto leaves of laurel or the like placed underneath, so that later it may flow off into another pot, and this is the *capitellum*. Now, if you want to make soap, put in a second water following the first, and when that has run down, put in also a third, and it will be good until it becomes white. Afterwards, melt some tallow, strain it, and when it is strained and cleaned on top if necessary, boil it with the last water. When it becomes thick, put in some of the second water and also some of the first. Or else if you soak ground poplar berries for a day [in the mixed lye and tallow] and afterwards squeeze and discard them, the soap will be reddish and better. This is French soap and *spaterenta*, i.e., sharp.

288-E. [A multiplication table]

i	ii	iii	iiii	v
ii	iiii	vi	viii	x
iii	vi	ix	xii	xv
iiii	viii	xii	xvi	xx
v	x	xv	xx	xv[201]
vi	xii	xviii	xxiiii	xxx
vii	xiiii	xxi	xxviii	xxxv
viii	xvi	xxiiii	xxxii	xl
ix	xviii	xxvii	xxxvi	xlv
x	xx	xxx	xl	l

[199] The translations lose the pleasant sound of the Latin words (*Decunx, Quadrans, Quincunx, Sextans*, etc.), which are references to the number of twelfth parts of the pound. These symbols are not used in the body of the manuscript, for which see footnote to chap. 1, p. 28.

[200] Symbol misplaced; see comment under "Z" in Appendix C.

[201] *Sic.* A later hand—perhaps Phillipps's—has corrected this to xxv. The lower parts of the first three columns are congested to the point of near unintelligibility. Berthelot, with misplaced ingenuity, supposed that this table (which is almost the earliest of its kind known) recorded musical notes and overtones in acoustic pipes.

288-F. [*An alloy*][202]

Melt together 9 oz. of tin, 2 oz. of copper, 6 oz. of silver.

288-G. [*A trick fountain*]

By means of the figure *arragab*, the fountain will play or stop as you wish.[203]

288-H. [*A trick goblet*]

By the same [arrangement] a goblet will either let the drink flow out or retain it.

288-I. [*Another trick fountain*]

If oxen drink first out of figure A [in a fountain, the diagram of which is missing], there will be plenty for both the oxen and the horses. If horses [try to] drink first, there will be none for either the oxen or the horses.

---

[202] This alloy (53 per cent Sn, 35 per cent Ag, 12 per cent Cu) is too rich in tin to cast well or to take a good polish as a mirror, and is too rich in copper and silver to be a good pewter. It might perhaps be intended as a solder of intermediate melting point, or as a hardener for pewter.

As Phillipps reported, the leaf preceding this one has been torn out, leaving only a small fragment of its lower corner attached to the binding between folios 64 and 65. It bears the single letter *s.* on what had been the verso. Did it perhaps contain the continuation of the multiplication table, torn out by an eager user?

[203] This and the following six chapters, which in the manuscript run together without division, are brief excerpts, essentially only the titles, of descriptions of various pneumatic devices and automata of the kind described by Philo of Byzantium and elaborated by Hero of Alexandria. Except for the constant-level fountain they do not, however, coincide with any of the devices described in the published manuscripts of these authors' works. See Wilhelm Schmidt (1899). The most complete version of Philo's works is in an Arabic manuscript in Istanbul published with French translation by Carra de Vaux (1903). It is not unlikely that the chapters in the *Mappa* were retranslations from a similar Arabic source omitting the diagrams but not the references to them in the text. Philo's original diagrams are labeled with Greek letters in the usual alphabetical sequence, $\alpha\beta\gamma\delta\epsilon$. The word *arragab* in the second sentence is hard to decipher. Berthelot (who missed the nature of the devices and thought that they were a collection of magic spells) suggested that this was a corruption of *arrazgas*, lead (see note to chap. 195), applied to a cast figurine. To us, it seems far more likely that the figure referred to is a diagram, and that *arragab* is a corruption, through successive transliteration from Greek to Latin to Arabic and back to Latin, of letters originally used to designate points on illustrations that have not been copied. (The loss of the drawings is probably not serious, for Philo's and Hero's drawings suffered even greater increases in entropy at the hands of copyists than did their texts.) There is not enough information to reconstruct the mechanisms, though the effect aimed at is usually clear enough. The devices in chapters 288-G to 288-M as well as the Cardan suspension of chapter 288-O seem rather more elaborate than the other gadgets that Philo describes, but they are not beyond his capacities or interests. They may, of course, simply be later additions by an ingenious follower. The material in the adjacent chapters is not in any way related to the Philonic devices, and is, in fact, unusually diverse in regional flavor.

288-J. [*A wine fountain with a constant level*]

From the same, wine will run out from a large jar into a basin until it fills the basin; when the basin is filled, no more will run out of the jar.

288-K. [*An apparition*]

Similarly, in an oil-burning lantern, by means of sand and a pivot and water, an apparition will come out of a tiny house and go in again.

288-L. [*Corydon*]

When the things below have been closed by means of fire and water, Corydon will winnow [his grain].

288-M. [*A toy castle with moving soldiers*]

When through figure A the spear [which acts as a plug] is removed, soldiers will come from the castle and will return as the spear hisses.

288-N. [*Mortar for a well*]

Use an egg in lime [when] the lime [is used in mortar] in a well.

288-O. [*Gymbals*]

When the lower four circles turn within each other, the arrangement of their diameters being as set forth [in a missing diagram], and a pot suspended in the center, nothing will pour out whichever way the circles turn.[204]

---

[204] This unspillable pot is set in a universal joint of the type later to be called the Cardan suspension, after the sixteenth-century author who described its use as a suspension for a magnetic compass. A nearly identical device is the polyhedral inkwell described and illustrated in the Arabic Philo manuscript in Istanbul referred to in the footnote to chapter 288-G. According to De Vaux's French translation, the device can be octagonal, hexagonal, square or "any one of the shapes that are given to prismatic glasses." There is a hole in each face, and one can dip in the pen through whichever hole happens to be on top and write. "In the interior there is a ring on an axle $\alpha\beta$; within this ring is another one on the axle $\gamma\delta$; in the center of the second ring there is a little cup on the axle $\epsilon\zeta$ and it is this that forms the inkwell. If you please, this is in the Jewish manner and the construction resembles that of a censer which turns and remains always in equilibrium. . . ." Though the principle is the same as the device in *P* the actual account is clearly independent. It is not impossible that the device was described by Philo, though the references to the prismatic glasses and the Jewish manner are both obviously later. Joseph Needham (1965: pp. 228–236) believes the device originated in the Far East. He cites a Chinese description of the second century A.D., and illustrates several from the eight century and later. A fine early European one is the sixteenth-century copper handwarmer (incense burner?) at St. Riquier, illustrated as No. 77 in the catalog to the exhibition *Les Trésors des Eglises de France* (Paris, 1965).

288-P. [More runes][205]

ᚱ	a	ᛁ	i	ᚱ	r	
ᛒ	b	ᚷ	k	ᛋ	s	
ᚻ	c	ᚱ	l	ᛏ	t	
ᚦ	d	ᛗ	m	ᚢ	u	
ᛗ	e	ᛏ	n	ᛞ	x	
ᛈ	f	ᛜ	o	ᛦ	y	
ᚸ	g	ᛈ	p	ᛉ	z	
ᛉ	h	ᚳ	q			

288-Q. [The Greek alphabet][206]

ᾳᴧᴧ	[A α]	alpha	a	Νᴪ	[N ν]	[nu	[n]
ʙu	[B β]	beta or vita	b	ᛦᛦᛦ	[Ξ ξ]	xi	[x]
ᴦᴦ	[Γ γ]	gamma	g	O	[O o]	[o]micron	[o]
ᴕᴧ	[Δ δ]	delta	d	ᴄᴑᴫ	[Π π]	pi	[p]
∈ᴕ	[E ε]	epsilon	e	ᴑᴘ	[P ρ]	rho	[r]
ᴣ	[Z ζ]	zeta	z	ᴦ	[Σ σ]	[sig]ma	[s]
ʜᴵʊ	[H η]	eta	long e	ᴛᴛ	[T τ]	tau	[t]
ᴕᴕ	[Θ θ]	[t]heta	th	ᴠy	[Υ υ]	uui[upsilon]	[u]
ᴵ	[I ι]	iota	i	ᴓ	[Φ φ]	phi	f
ᴋᴌ	[K κ]	kappa	[k]	ᴕ✝	[X χ]	chi	ch
ᴧᴦ	[Λ λ]	lambda	[l]	ᴪᴪ	[Ψ ψ]	psi	ps
ᴧᴧᴧ	[M μ]	mu	[m]	ᴜ ᴕᴑ	[Ω ω]	omega	long o

---

[205] See also chapter 288-B and Appendix C.

[206] The scribe clearly did not understand what he was copying, for he omitted a whole line of Latin equivalent letters and erroneously aligned the entries for theta, iota, sigma and tau in the margin. Theta and sigma have been partly mutilated by a binder's plow

289. To cut glass

With a stinging nettle the Saracens stingingly nettle the udders of a she-goat, and thump them with the palms of their hands so that the milk descends into them. The milk is subsequently milked into a vessel in which the glass is put overnight, together with the iron tool with which it is to be cut. The tool will be hardened [by heating and quenching] in the milk, or in a small red-headed girl's urine that has been collected before sunrise. Now, when necessary, the milk should be re-heated to the same temperature that it had when it was freshly milked, and the glass should always be heated in it until it becomes soft and so it may be cut. Other gems also in the same way. The goat, now, should be fed on ivy.

290. To chip crystal into shape

Take a he-goat which has never copulated and put it in a cask for three days until it has evacuated everything that it has in its stomach. Then give it ivy to eat for four days. Then clean a jar for collecting its urine. After this, kill the goat and mix its blood with the urine, then put the gem stone into it for a night, and after this, either chip[207] it into shape or engrave it if you wish. To make it beautiful, make yourself a lead slab and sprinkle on it white flint, ground like pepper, and rub the stone on it until you smooth the roughness away. Afterwards wrap up some of the same ground flint in a woolen cloth, and with it rub the corners that you were previously not able to shape on the flat sheet [of lead]. Then, so that it may recover its pristine clarity, make yourself some oil from nuts and rub it with this. Further, you should smear it over with a waxed cloth, so that it becomes brilliant and ceases to sweat.

291. Gilding iron

Copper filings are ground in a bronze mortar with vinegar, salt, and alum to the consistency of honey. Some people use water instead of vinegar. Then the iron is well cleaned and gently heated, coated with this mixture, and is rubbed until it takes on the color of copper. After this, it is washed off with water,

and wiped dry. Then it is gilded in the same way as copper or silver, and heated to drive off the quicksilver as usual. It should be rubbed with a [burnishing] tool so that it acquires brilliancy.[208]

292. Another way

Rounded and well-preserved alum, which is called rock salt,[209] blue vitriol and some very sharp vinegar are ground in a bronze mortar. The cleaned iron is rubbed with this [composition] using a stick or some other smooth little point. And when it has taken on the color of copper, it is wiped again and gilded, and then, after the quicksilver has evaporated, it should be cooled in water and rubbed with a tool that is very smooth and bright until it becomes brilliant.

293. Ivory

Now, if you want to straighten ivory or to bend it, it should be put in this above-mentioned mixture[210] for three days and nights. When you have done this, hollow out a piece of wood in whatever way you want, then put the ivory in the hollow and straighten or bend it as you wish.

293-A. [A wall plaster]

Take 2 parts of quicklime, 1 part of ground tile, 1 part of olive oil, one part of chopped tow: mix all these with a lye made from elm bark.

294. [Some identifications][211]

Cinnabar	i.e.	vermilion
Iarin	i.e.	efflorescence of copper
Psimithium	i.e.	efflorescence of lead
Mag[nes]	i.e.	sinopia or Armenian bole

---

[207] *Comprimo*, literally compress.

[208] This and the following chapter on the same subject are nearly identical with chapters 146-G and 146-H, which also illogically follow a discussion of the polishing of gems.

[209] In chapter 146-H, which is nearly identical with this, the materials called for in the first sentence are, more plausibly, "Rounded alum, the salt that is called rock salt, blue vitriol and some very sharp vinegar."

[210] Perhaps the mixed goat's milk or urine of chaps. 289 and 290?

[211] These are written in a different but nearly contemporary hand on the back of the last leaf of the manuscript.

APPENDIX A

THE SÉLESTAT MANUSCRIPT
MS 17
in the Bibliothèque de la Ville
de Sélestat, reproduced from a
microfilm prepared by M. Grelot
of the Archives Municipales,
Strasbourg.

In the original the text is approximately
13.6 cm. high, measuring from the top of
the first line on folio 2 to the bottom of the
last line. In this reproduction only the
recto of each leaf is numbered, the *verso*
being mounted directly beneath.

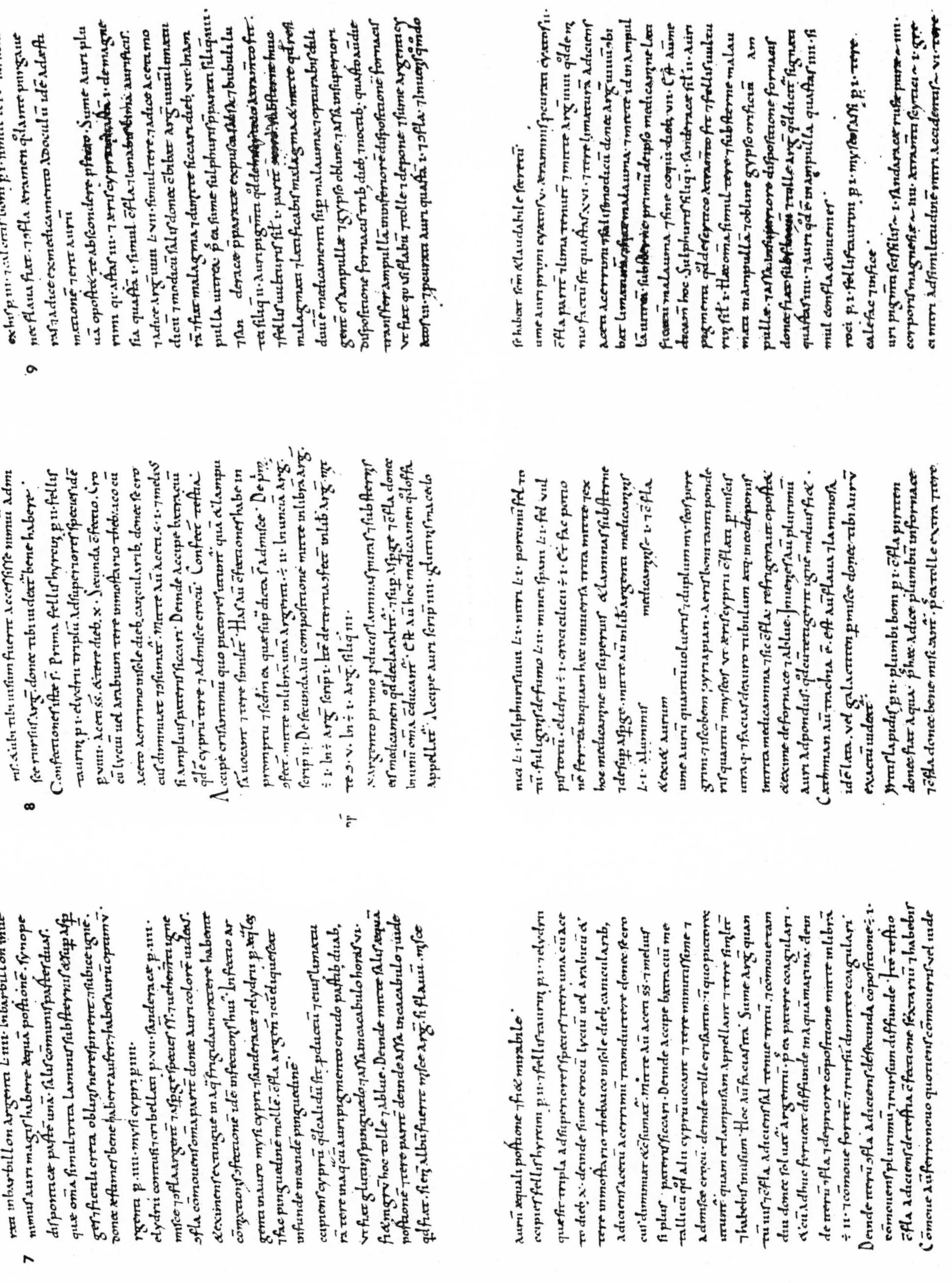

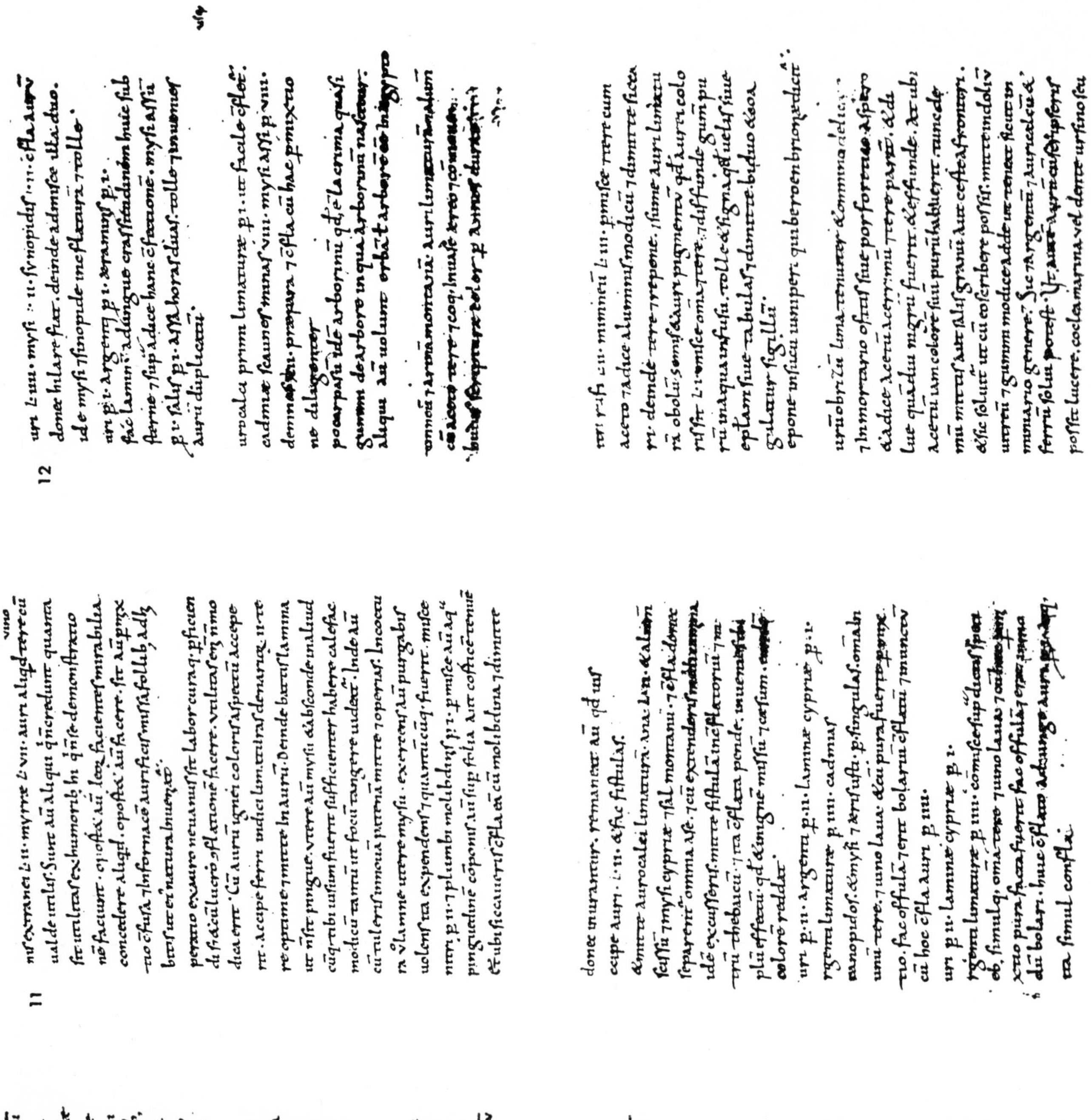

14 DISPOSITIO FABRICAE DEPONENTIBVS VEL QVIB

mensuris oportet aedificia disponere vel quib. mensuris
mensuris oportet secundum modum fabricae. Si in altitudine quaeris
saxum fuerit fabricae unius. facturus altitudine. oportet
te fundamentum. Si uero trib. saxum. altitudo usque ad h.
saxum fundamenti. Si autem unius facturae altitudo usque...

DE FABRICA IN AQVA

Si fabrica in aqua necesse fuerit erigere. faciaris ex
angulis. aequis a fonte custodi disponet. ut in medio necesse...

Fola floris uole colliget & in mortario mundo teret bene &
mittet sapone euacuans...

Me fuerit... hormis ipsa... ut remov...
nunc imponet lapide adsafabricanda. Temperatio in aliis...

De compositione cynnabrii

De compositione cynnabrii

Tolles... mundu... uinis parte una...

Compositio plumbi

Tolles... plumbi & fariet pilulas suspende sup. acetum & colliget...

---

13 Lapide onichino litterae & fabrica.

...urispigmentum p. iiii. auri p. i. sal. auri p. i. deduce...
...urispigmentum cui color sit auri...

...cannam si nudo alumine & tunc in aurum utrunque...
...scribe. & siccet inde inde. & sic diligenter...
lumen fac nihil. in aqua. deco. flume argenteum...

16

17

18

[Leaf 16 is a small interpolation correcting the omission at the head of 15 *verso*. 16 *verso* is blank.]

82

22 Compositio auri...

Compositio electri

Inauratio vitri lapidis et ligni

Crisografia de petalis

23

Compositio litargiri

Alia compositio litargiri

Inauratio argenti examinati et auri...

24

Compositio brundisii

Compositio afronitri

Quomodo fiat sulphur coctum

Compositio litri

84

25

26

27

TINCTIO MVSIVI PRASINON.

DE COLORATIONE MVSIVI.

DECOCTIO PLVMBI.

DE METALLO VITRI ÆCOCTIONE.

FROBRITIO LÁ MEALLI.

28 inllud. & dimitte dieb, v. Postea eice & laua.

De uenai tinctione eorundem.

I puenetu aū tinctione aluminē quod uis eorū sic p̄dixim⁹.
& facto lulacterim. mitte inllud si ossus fuerit dieb, x. Si uero
cornu dieb, viii. Si aū lignum fuerit. diebus tribus.

De eorundem tinctione melina:

I ptinctione ꝰ melina. alumina que tingenda sunt sic sup̄
diximˀ. & coque luzam cū urina dispumata. & cū bullierit
depone in illā. De colore cynnabri.

S i colorem similē cynnabrin facere uolueris. sume sinopidē coctā
partes duas. Siricum partē unā. cōmisce & tempera cum
aqua. De pergamena bouina.

P ergamenam bouinā si facere uolueris. mitte pellē in calcē. &
iaceat ibi dieb, tribus. p̄ea tende illā in cantiro. & rade eā p nouacula
exutraq. parte. & mlam. siccare p̄mitte. Deinde quicquid
uolueris scappillatura facere. & postea pinge coloribus.

S ume ac&ū acerrimū. & funde Cōpositio psimithin.
in amfum usq. addimidiū. Deinde laminas plumbeas tenues
suspende sup ac&ū. ita ut humore n̄ tangant. & obturato
diligenter ore amfine habeat spiraculū. Pone eū t ad solē

t in alio quolib& loco calido. ubi immobilit̄ stare ualeat. Et p
dies. xv. t (aut) xx. pro amixto collige florē qui nat' ē in p&alis.
& lotū aqua ad solē desicca.

C ebellinum ta fit. tolle lignū cerrinū aut derrinum. & mundi
ramos ei' excortice. & facitē ei' leuit̄ dola. deoperi illud in loco
cenoso annis. xx. Postea eiectū de ceno. dimitte in umbra sicca
re anno uno. & labora exeo quod uis.

De specieb. m&allorū. herbarū. lignorū. lapidū. atq,
fugi salis. nitri. afronitri. olei. picis. rasinæ. sulphuris.

P rimū m&allū exquo fit aurū. t̄ra rufa ammoydis subru
bicunda. propter aduix fiantem illi t̄ram. Est enī & alia
similis. & dū incendit̄ pd& colorem. & n̄ ē arenosa sic illa prior.

N ascit̄ aū in solanis locis ea t̄ra. & tale ē m&allū auri. M&allū
uero argenti uiride ē. M&allū aū aramenti. p&ra ē uiridis.
quā dū p artis cū pyregliolo ignē emittit. Auricalci aū pe
tra ē melina. & eodē modo ignē emittit. M&allū lapis ē co
lore ipso gatizom. Plumbū aū t̄ra ē fuscā. Lapis uero que
in ea inuenit̄ uiridis ē. Arena ē unde uitrū m&allizatur.
ē enī & ipse lapis uitrei coloris. Vitriolū uiride fit t̄re.
ogri zossunt cr&e ubi uerno temp̄ guttant. Ipsā colligent

29 & decoquent. & ipsa terra fit calcitarin. que aū arida uitriolū.
A luminis aū m&allū ē t̄ra floriens. Ertarin terra ē alba. faci
lis adpiscandū. Sulphur extra nascit̄. & ipse incendit̄ locus.
coctū aū extra sulphurea oleo mixta coquit̄. Nitrum salem
quod nascitur in terra fit in lumnas in temp̄ caust̄. Sal scistis
nascit̄ similit̄. Afronitrū uō nascit̄ in lonitri priusquā gel&.
Cōponit̄ aū & aliud ex nitro. Principale aū spuma alba ut nix.
cōpositū uō plus fuscū ē. hab& tam eandē uirtutē. Terra
sulphuritana. nascit̄ in eodē loco. ubi sulphur nascit̄. Ipsa enī
t̄ra generat sulphur. Lapis emaitites. inuenit̄ iuxta locū
ubi sulphur nascit̄. Argentū uiuū nascit̄ ex terra. nascitur
& aliud ex m&allo argenti in conflatione. Auripigmentum
m&allū ē terre. Calc&is gleba est naturis que in cypro in
sula inuenit̄ in m&allicis colore subauroso. intus uenas habens
desistas. & alium ē scissum. & in modū stellarū fulgentes. Pra
sinus t̄ra ē m&allizat. Lulax cōponit̄ ex terra & herbis.
Lazurin compositū ē. Cyanus cōpositus. Ficarin compositū.
I arin æraminis ē flos. Psimithin plumbi ē flos. Ocria terra
ē pandia. omnia colores omnes cōpositio. Calcucecaumenū
ex æramine fit. Cynnabrin ex argento uiuo fit. Sericum

fit ex psimithio. fit & iā ex plumbo. Herbarū aū t̄re & lignorū.
C risocollon arbor ē p alta. melinum habens interiorem lignum.
Pucis cortices & frutices. Olicine cortices. Melia cortices. Vlmi
cortices. Celsa cortices. Hec omā tinctiones sunt. Rubia saluatica
lita est. Monaclosus galla glandis ē. Tamusati carpustris ē
anticarbus. Trimalin herba ē. Dranta lasis dissobo gauzo herba ē.
R esina omni specie ex pino & sappino coquitur. Picero cocerta
petola semel. Cedria ex ligno cedrino coquit̄. Resina & sapino
nascit̄. Resina abietina ex abi&e nascit̄. Mastice ex lentico nascit̄.
Z igea extingueo. Gumma ex atrinia. Scita gumma ex amigdal-
O leū ex oliua. Lineleon ex semine lini fit. Oleū lentiscinium
ex lentisco fit. Colaliam ex mare. Cōnquilium ex mare. Sal
ex mare. De inauratione ferri.

S i aū ferrū deaurare uolueris. sume calcitarin & alumen asianū.
æquab, ponderibus. salis aū & dracanti tantā quantitatē que
compens& tota tria. & cōmisce hæc omā cū aqua t ac&o acerri
mo. mittens ea in cacabū æneū ut bulliunt una hora diei. Postea
extersam & pumicatā diligent̄ ferrū cū hac cōfectione p unguē
in eo loco ubi illud deaurare uolueris. Et cū modicū ita ee in
hanc unctione ee patieris. terge illud. & habebit colorē æneum.

30 D einde sume lapide onichinū & liza illud. & similizando colorē
pdiderit. iterū tingue. Si uō in auratura p̄hendere noluerit.
cōmisce cū medicamine qualiter & inungue sic sup̄ dixim'
de blatta tauratica inqualem uis pannū facere uolueris. & si
lixzandū ē. cū onichino liza. Adp&alum aureum.

L ineleon calbani ÷ ii. Terrebentine ÷ i. Pice spana ÷ i. Istas tres
species. idest galbanū. terrebentinā. & picem. solue in unū cū
modico lineleon. Postea uō cum orientalem ÷ ii. Libani ÷ iii.
Mirra ÷ ii. Mastice ÷ ii. Resina sappini ÷ ii. Floris papuli pri
motici ÷ ii. Veronice ÷ ii. Lineleon & uricella cōmisce &
ad mosana colas. Postquā tota fersa fuerint. misce ibi de
gumma cerasi ÷ i. Hec omā in unū mixta. bullire facias in
lineleon m̄sura ÷ iii. uitriarū. Post coctionē aū cola p linteū.
& misce species sup̄ dictas. idest galbanū. terebentinā. & picē spanā.
Et si aliquod uitiū p̄ea habuerit quo siccari n̄ possit. addito de
mastice quantū uolueris. aut unciā scilic& aut dimidiā.

Lucida sup colores ducenda. qualit̄ fieri debeat.

L ineleon ÷ iii. Terrebentina ÷ iii. Galbani ÷ ii. Lirice ÷ ii. Libani ÷ iii.
Mirre ÷ iii. Mastice ÷ iii. Veronice ÷ i. Gumma cerasi ÷ ii. Floris po-
puli ÷ ii. Gumma amigdalina ÷ iii. Resina sapini ÷ ii. Hec omā

pisanda t & cribellanda. & cū sup̄ dicto lineleon ponenda in
gabata auricalca. mittendaq. in furnū calidū. ibi sine flāma
coquant̄ eodē modo ut foras n̄ exeant. Cocta aū colanda cū
linteo mundo. & si nda uenerunt. decoque ea usq, dū spissa
fiant. Postea uō quolib& opera t picta t scapellata uolueris
inlucidare poteris. Lucidata autem siccare potueris.

A urum obrizum lima tenui. De crisografia.
lima. & in mortariū porfor iticum mitte. adiciens ac&ū
acerrimū. & terens parit̄ & lauas quādiu nigrū fuerit &
effundes. tunc demū mittes aut salis granū aut certe nitrū
& sic soluitur. Postea scribes & litteras polies. Sic omā m&alla
soluuntur. Item alia.

A lio aū modicum confla plumbū frequent̄ & r&ingue semp
in aqua frigida. Deinde conflatū aurū r&ingue. Postea le
m&tū tere diligent̄ aurū cū argento uiuo. & purgatū tem
pera cum liquido gummi & scribe. Ante aū in alumine le
quido galomum tingue. quod alumen cum sale & ac&o
optimo purges. De litteris poll.

S ume laminas aureas & argenteas. & tere in mortario nitro do
nec non pereat. Deinde mittes aquā & infundes. & iterum

DE INAVRATIONE PELLIS.

DE AES ALBO.

DE GEMMIS.

DE BASILISCE.

De calce & harena.

De columnis.

87

40

[Leaves 35 to 39 are omitted in this reproduction as they contain nothing related to the *Mappae Clavicula*.]

39v

34

The lower part of leaf 40 *verso* is blank. A large rectangular piece has been cut cleanly from the upper part of leaf 41: It would have contained chapters 264, 265, and the latter part of 266.

41 S.

[A portion of leaf 41 has been excised].

S

utiliter. De sagitta que ignem emittit.
Sagitta ad emittendum ignem. Tribulus est pertusus.
Confectio autem talis est ignis. Napta ÷ I. Stupio solidos
II. Pice cedria solid. IIII. Sulphuris uiui munda guttae
solid. I5. Climatidos solid. I5. Salis marini solid. I.
Oleum exoliuis solid. I. Visco aspero solid. I. Lapidis
gagatis solid. I. Saponis ex oleo solid. I. Ista tol
lens et mittens in mortario marmoreo cum tritorio.

c. ig. Alia sagitta que ex modica compositione ignem emittit
Sulphur uiuum. colofonia. resina aequis ponderibus.
Oleum ex nuce conficies et mittes sic supra diximus.

Sagitta in toxo balestra uenenata ut non incendat canale.
esto ipsum canale excere. In sagitta uero tribulos secundum priorem
compositionem utere. et petra focaria que ignem emittit
pertusa et inaequalis ubi ipsa compositio possit contineri.

42 Drapidos compositio ex sulphure lib. I. Napta lib. I.
Stupium lib. I. Clymatida lib. I. Pice condita lib. I. Lac
muliebre ÷ II. Olei porcini ÷ I. Lapidis gagatis ÷ I.
Resina ÷ I. Resina condita lib. I. Sulphuris uiui lib. I.
Pice liquida ÷ I5. Cedrea ÷ I. Oleum exoliuis ÷ I.
Sulphur coctum ÷ I. Auripimenti ÷ III. Nitrum ÷ 5 .T.
Omnia collecta arida diutius tere. et post haec tollens
omnia humida commisces. tere utiliter. et compositis utere.
pungens petram. et omnes pertusas eius replens. et imponens
in petraria. Immittes ignem. et dimittes celeriter.

Compositio arietis ad muros expugnandos. Anteriores pe
des tres cubitorum V. Dimidio cubitorum IIII. Retrorsus cu
bitorum III. Rotae autem altae unius semis palmi. grossae
÷ IIII. Circinas et in medio pertundes. secas columnas. et immit
tis rotis. usque ad IIII. unciarum. Cooperies rotam et super conexi
onem facies. et confige cum meura. Astringes arietes. et con
texes funibus. pertegeris cum corio. et super filtris cooperies. et super
filtra coria et super coria arenam ÷ IIII. et super arenam lanam. ut
permoueat ipsa arena. et desuper coria. Tales autem habeant

ipsae columnae cardines. ut dum ambulat conuertas
ubi uolueris. si delaxas cardines permouetur. qui confi
guntur intus. Et rotis subpositoria subpones. et ipso
ingenio coniungas muro. et labores indubitaviter.

Quomodo debeas zelum arietis incendere. Compones cac
cabum percoctum compositione dedimus. Ignem appones.
et in ipso zelo emittes et incendes ipsa coria et lanam.
post haec remanet arena et ipsa confixa eo quod perardeat.
Post haec de lapides permoles ipsam arenam. Compones iterum
similem caccabum. et ex ipsa compositione iactas in ipsum zelum.
et si propter multitudinem arenae perardeat. iacta alios lapi
des. et compone alium caccabum et iacta.

Compositio dedimus electri pugnatoris. Compositio
artimilli et idoruthora cum macu bellerofantu per
seunos cyronon elandi. onos milionos. In compositione
alemtiu gentio nossantonio. et abrochiu. aquilino.
Haec compositio electro similis. Compositio IIII. formas spe
cierum. Nathopisso. Stupium. Clymatidos. Compositio

43 naptae haec est. I. Naptae mundissimae lib. I. II. Sulphu
ris uiui lib. II. Milino ÷ IIII. III. Salis marini penso solid. IIII.
IIII. Oleum de oliuis mox collectis et macinatis. uel unius duorum
annorum pensum solid. IIII. V. Colofoniae mundae liquidae
÷ II. solid. I. VI. Peculae solid. IIII. Aureas VI. VII.
Picis durae ÷ I. VIII. Olei terebintini dracontoides ÷ I.
Compositio autem olei terebintini haec est. Olei terebintini
partem I. Olei communis partem I. Olei laurini partem I. sol. II.
VIIII. Cedreae cedrinae sol. II. X. Picis cypressinae sol. II.
XI. Mastice sol. I. XII. Lapis gagatis. Haec est compositio
naptae XII. specierum. Coctio autem talis est. Teres omnia
arida. humida uero commisces et ipsa defricas. Dum autem con
teres omnia arida utiliter. et defricata omnia humida co
misces cum aridis et defricas omnia bene. Post haec repones
in uaso testeo bituminato. et dimittes duos dies aut III.
Et post haec tollens calefacies modicum ut bulliat. et dimittes
residere. Haec est compositio naptae.

Compositio picis.

Picis aridae ÷ I. et I. II. Sulphuris uiui et I. III. Re
sinae et I. IIII. Isata. Alii dicunt oleum aquae. Alii florem

aquae. Alii cicedonia. Alexandrini autem amorem aquae.
Nascitur enim in aqua ubi alba terra est. terra russea. terra
nigra. Exiente autem de aqua fit milinum super aquam incir
cuitu exitus aquarum. Est autem aqua grauis et egrota. pereter
que florem generat. Sic autem colligetur. A martio et aprile si est
calidus locus. usque ad octobrium mensem colligitur. Tolles autem lanam
mollissimam lotam. et impones super aquam. et exprimes in uaso vi
treo. habente pertusum modicissimum uelut acu factum. quem
cera clausum pones ad solem decem diebus ac noctibus immobilem.
Et post haec ablata cera. aperies ipsum pertusum. subposita lana
munda donec deliquetur aqua et remaneat amor aquae.
Itaque de amore aquae mittes ÷ et I. V. balsimi mundi ÷ I.
Aureas VI. VI. Oleum silicum. Alii oleum ricinum dicunt. Alii
laucidis. Alii uero uiscum sol. I. VII. Picis cypressinae solid. I.
Aureas XII. VIII. Picis pineae sol. I5. VIIII. Sapone ex
oleo sol. II. X. Nitri sol. I. XI. Selinifteo. Alii rodica.
Alii pauti. Alii rusticon. Alii gumma. Alii marticam.
Alii thymon. Alii trichas. Alii tricoselinon. Nascitur enim in
aqua uelut apium. et in parietibus ubi calx est. et dicitur herba
capillacia arida trita ÷ I. XII. Aolochias. Alii stacis.

COMPOSITIO ARPII

DE LAZIDE OLIMPIO

COMPOSITIO LULACIS

Purlit compositio

TINCTIO RUBEA

LAZVRIN ATIZONTA

ALIVS SANDIVS

LAZVRIN MILINIZONTA

LAZVRIN AVREI COLORIS

LAZVRIN OMICHRIZONTA

50 in letamine per dies xx. Item pandius ocrei coloris.

Ocra munda lib. i. cinnabarin ÷ i. sicarin sol. iii. Omnia tritas in mortario commisceas cum urina exspumata. et repones in vaso vitreo. et ponis ad solem donec desiccetur.

Item pandius purpureo colore positus ex iiii speciebus.

Lulacin. cianon. cinnabarin. laccha. aequis ponderibus trita et commixta ponis in vaso vitreo et reponis ad solem donec desiccetur. Item pandius porfirus.

Iotta decoctionis conciliii lib. i. cinnabarin lib. i. sirici mundi ÷ i. Omnia trita et cum modica urina commixta mitte in vas vitreum et repones ad solem donec desiccetur. Item pandius porfirus.

Iotta conquilii ÷ i. Iotta de laccha ÷ i. teres primum cinnabarin. et post haec commisces iotta conquilii. et iotta laccha et terens pones in vase vitreo ad solem donec desiccetur. Item pandius subporfirus.

Iotta conquilii ÷ i. cinnaberin ÷ i. croci ÷ i. iotta herbae lunae ÷ iiii. decoctae in urina omnia commixta penso sextern.

Item pandius.

Cinnabarin ÷ i. Iotta conquilii ÷ i. coctio rubie ÷ i. coctio finisci. teres primum cinnabarin semotim. post haec commisces omnia. et repones in vase vitreo ponesque ad solem donec desiccetur.

PANDIUS

Tolles iottam rubie et addis galle ÷ iii. terensque utiliter. tolles exiotta rubie libras ii. et mittes in vas vitreum cum ipsa galla trita. et dimittes per duos dies infundi. et post haec colas et addas exiotta conquilii ÷ i. tollesque utraque et mittas in vas testeum ubi debeat coqui. et addas caletarin ÷ i. cinnabarin sol. ii. et teres utrumque. et mittas ea cum supradictis rebus et decoques donec veniat ad tertiam partem.

Quianon. ualtalasion. Nascitur enim in locis humidis. nascitur quidem ex rore aestivo tempore. Colligitur autem sic. Tollens colliges illam. et repones ad solem donec desiccetur. et postquam desiccaverit teres bene. Et post haec tolles coclea maiore marina et lavans bene et terens. mittes ex ipsa coclea lib. xxx. de ualtalasion lib. iii. sapone lib. x. larin ÷ iii. Sapone autem ex oleo leviter cocto. Omnia commixta et trita repone in vase novo testeo. et operi in letamine. dimitte diebus sexaginta.

ITEM

Quianus autem nascitur sic per pensum ante commixtionem specierum marmorem tritum bene commisces secundum mensuram coctionis. Item pandius alius viridis.

51 Quiani lib. i. psimithin ÷ i. ista trita et commixta cum urina exspumata. Item pandius.

Crani lib. i. cinnabarin ÷ i. trita cum urina exspumata.

Haec omnia exposuimus. quamquam ex terrenis metallis floribus et herbis. Ita exponimus virtutes et operationes earum in parietibus et lignis linteolis et pellibus et omnibus pictorum. Ita memoravimus omnium operationes qui in parietibus simplices. in ligno cere commixtum suscepit lignum simplicem cum unctione collon commixtum. In pannum vero cere commixtis coloribus. In pellibus unctio collon commixtum.

De vii miraculis mundi manu hominum factis.

Quorum primum est. Capitolium romae salvatio civium maior quam civitas. ibique olim fuerunt omnium gentium a romanis captarum statuae. et deorum imagines. et in statuarum pectoribus nomina gentium scripta quae a romanis captae fuerant. et tintinnabula in collis earum appensa. Sacerdotes autem pervigiles diebus noctibus per vices ad harum custodiam curam habentes. intendebant si quelibet earum moveretur sonum mox faciente tintinnabulo. ut scirent quae gens romanis rebellaret. Hoc autem cognito romanis principibus verbo vel scripto nuntiabant ut scirent ad quam gentem reprimendam exercitum mox destinare deberent.

II. Pharus alexandrina super iiii cancros vitreos per passus xx. sub mari fundata est. Hoc namque mirum quomodo tam magni cancri fieri possent. vel quomodo deportari et non frangi valerent. qualiter fundamenta cementitia desuper herere potuerint. et quomodo sub aqua cementum stare valeat. et quare cancri non franguntur. et quare non lubricant desuper iacta fundamenta.

III In rhodo insula colossi imago aerea c.xxx.vi. pedum fusilis facta. Hoc mirum qualiter tam immensa potuit moles fundi. vel facta erigi ut staret. xv. pedum mensura altior est ista imago colossi. illa quae

IIII. romae est. Simulacrum bellerefontis ferreum cum arcu suo in smyrna civitate in aere suspensum nulla re innixum. sed magnetis lapidis qui habetur in arcu assumptione mirabiliter trahitur. et mensura temperata consistit. Est autem ponderis ei estimatio. circa v. milia librarum ferri.

V. Theatrum est in eraclea civitate marmoreum de monte insculptum ut omnes cellulae. mansiones. muri etiam et antra bestiarum ex uno solidoque lapide facta videantur. Et nemo in gyro ipso tam secrete solus vel cum alio loqui potest. ut non ab omnibus audiatur. qui in gyro ipsius aedificii consistunt.

213

[Leaves 51 verso through 212 recto have been omitted as they contain matter not related to the Mappa.]

214

212v

APPENDIX B

THE PHILLIPPS-CORNING MANUSCRIPT

As explained in the Introduction, the manuscript was damaged in the 1970 flood. This reproduction has been made from the clearest copy available, a xerox copy of a set of photo prints made from the original in 1962. A small amount of retouching has been done to rectify photographic defects, such as filling in the thicker parts of the bodies of the capitals, and eliminating dark stains at the edges of the page and in places where the ink had penetrated through from the opposite side of the vellum, but the text itself remains untouched, even though it is occasionally illegible. Some of the scribe's pounce marks remain visible. The arabic numerals that had been applied in pencil to designate the leaves and chapters rarely reproduced well and have therefore been eliminated entirely.

In the original, the height of the text on leaf 1, measuring from top of the first line to the bottom of the last, is 13.1 cm.

94

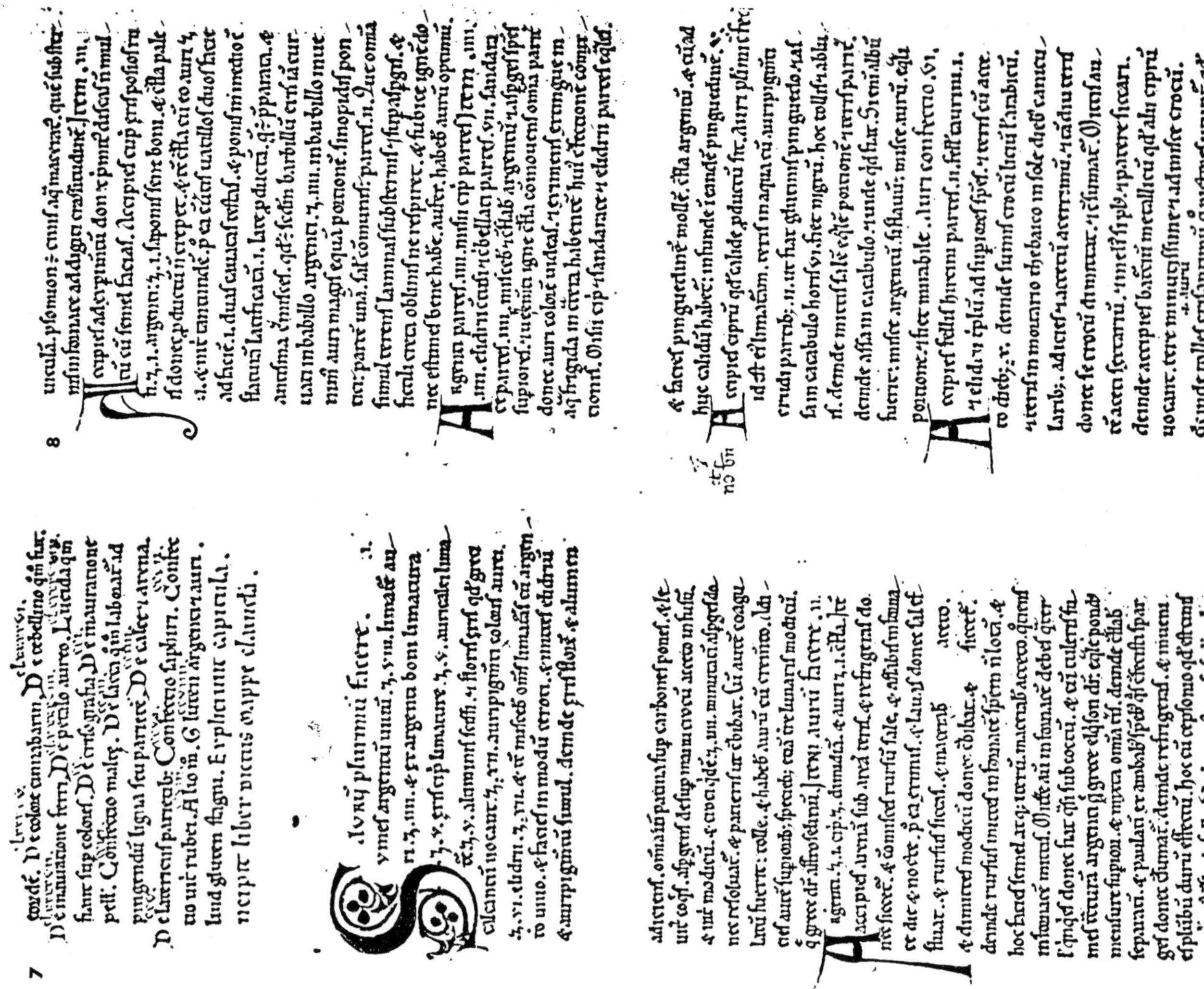

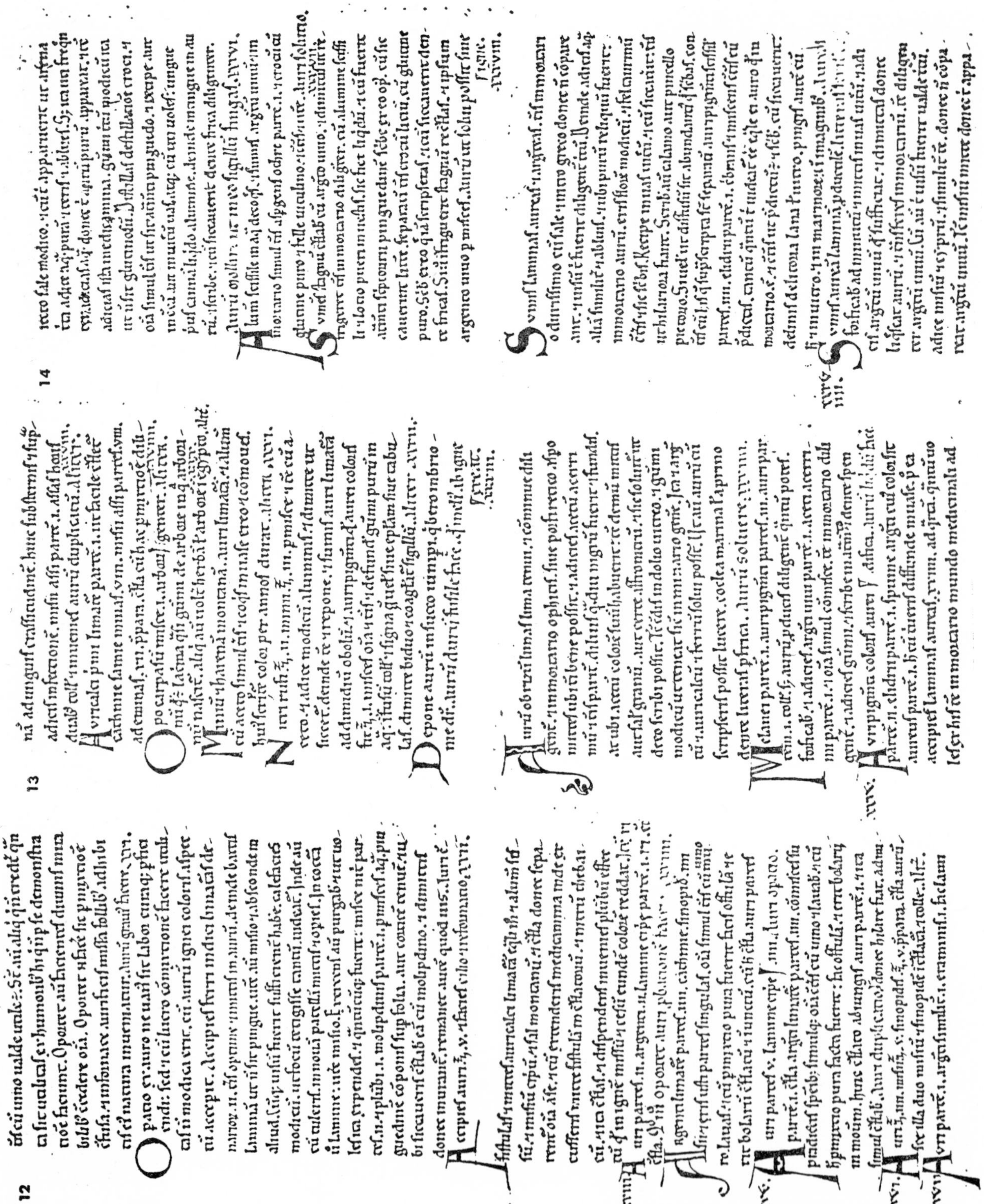

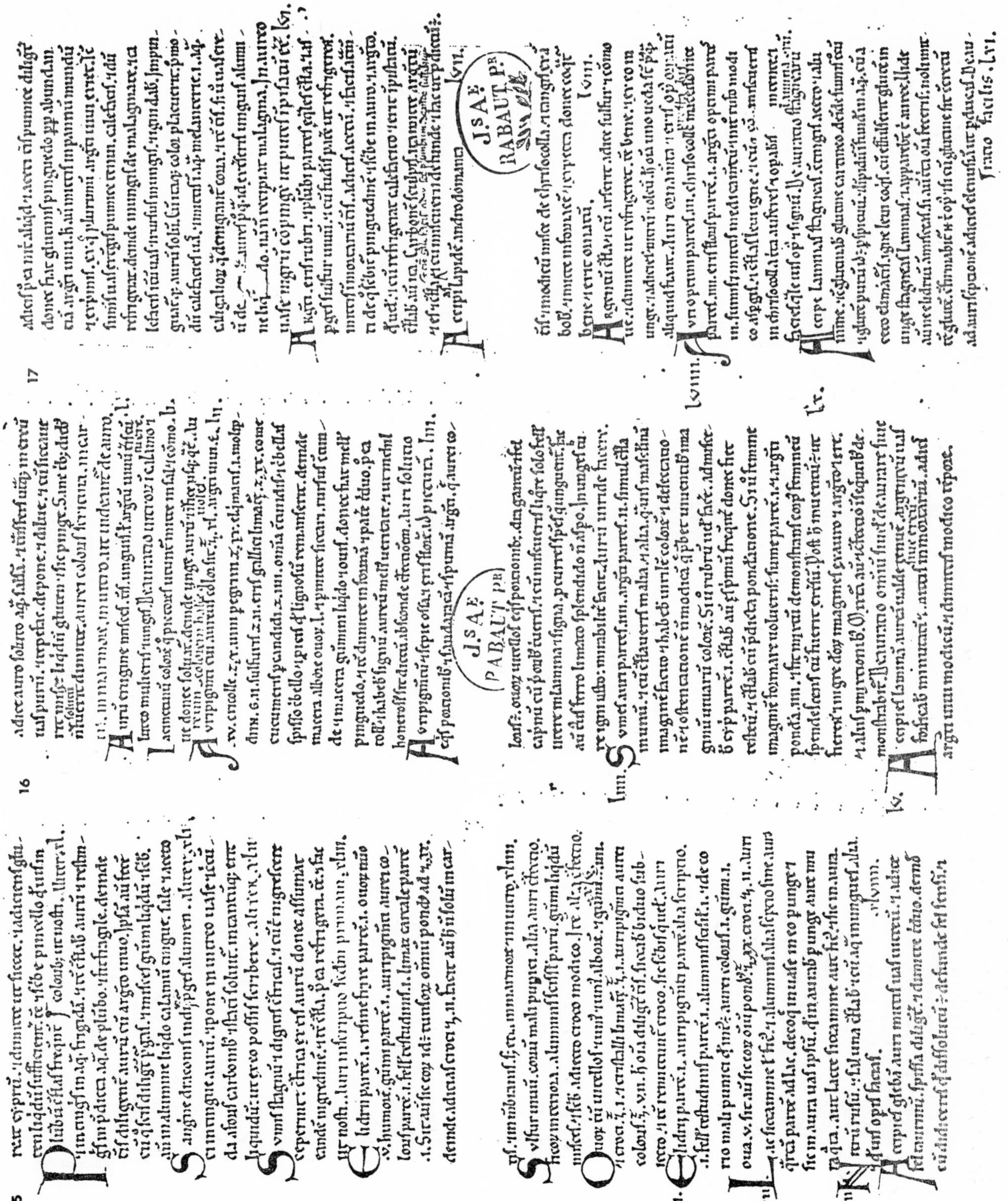

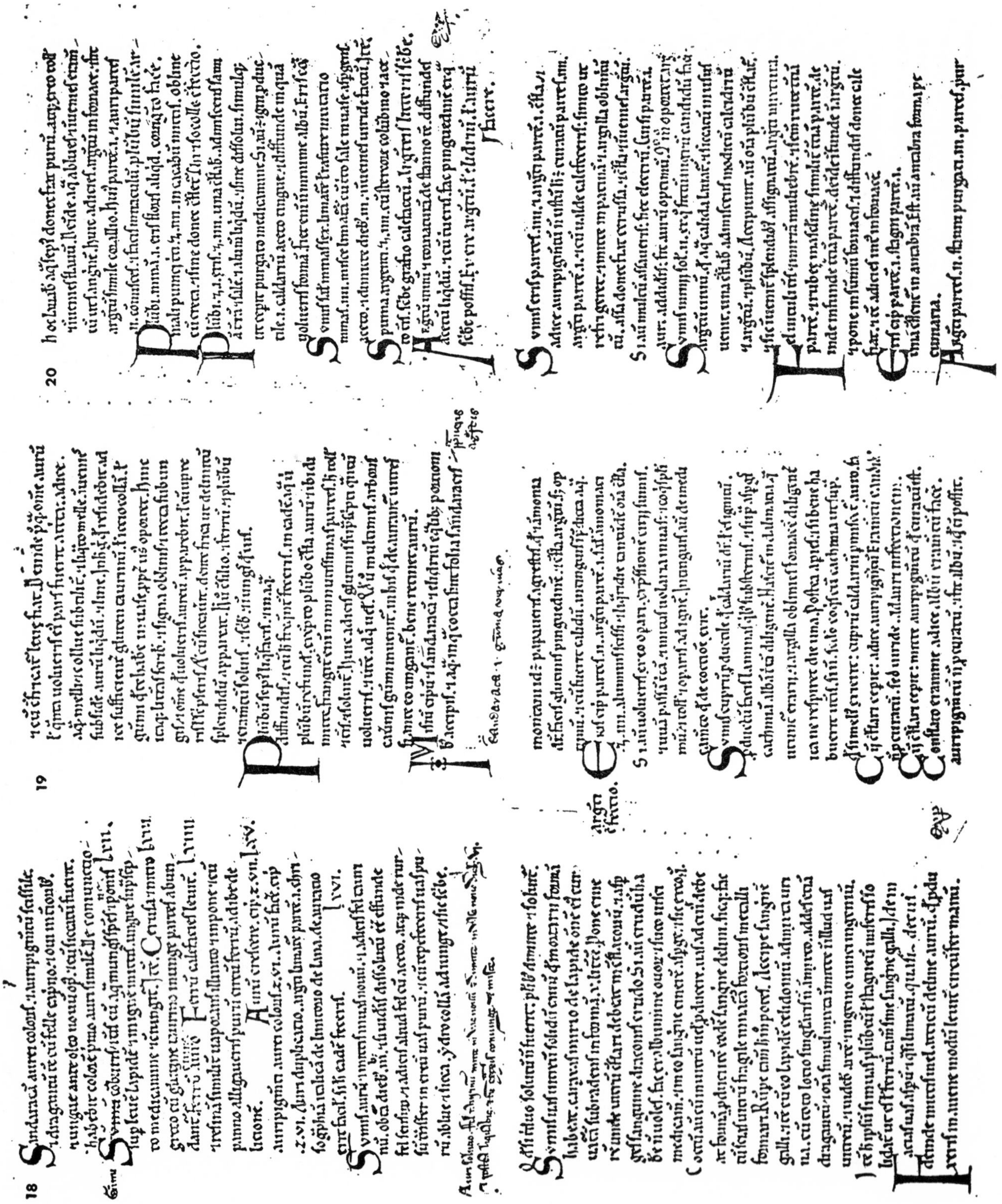

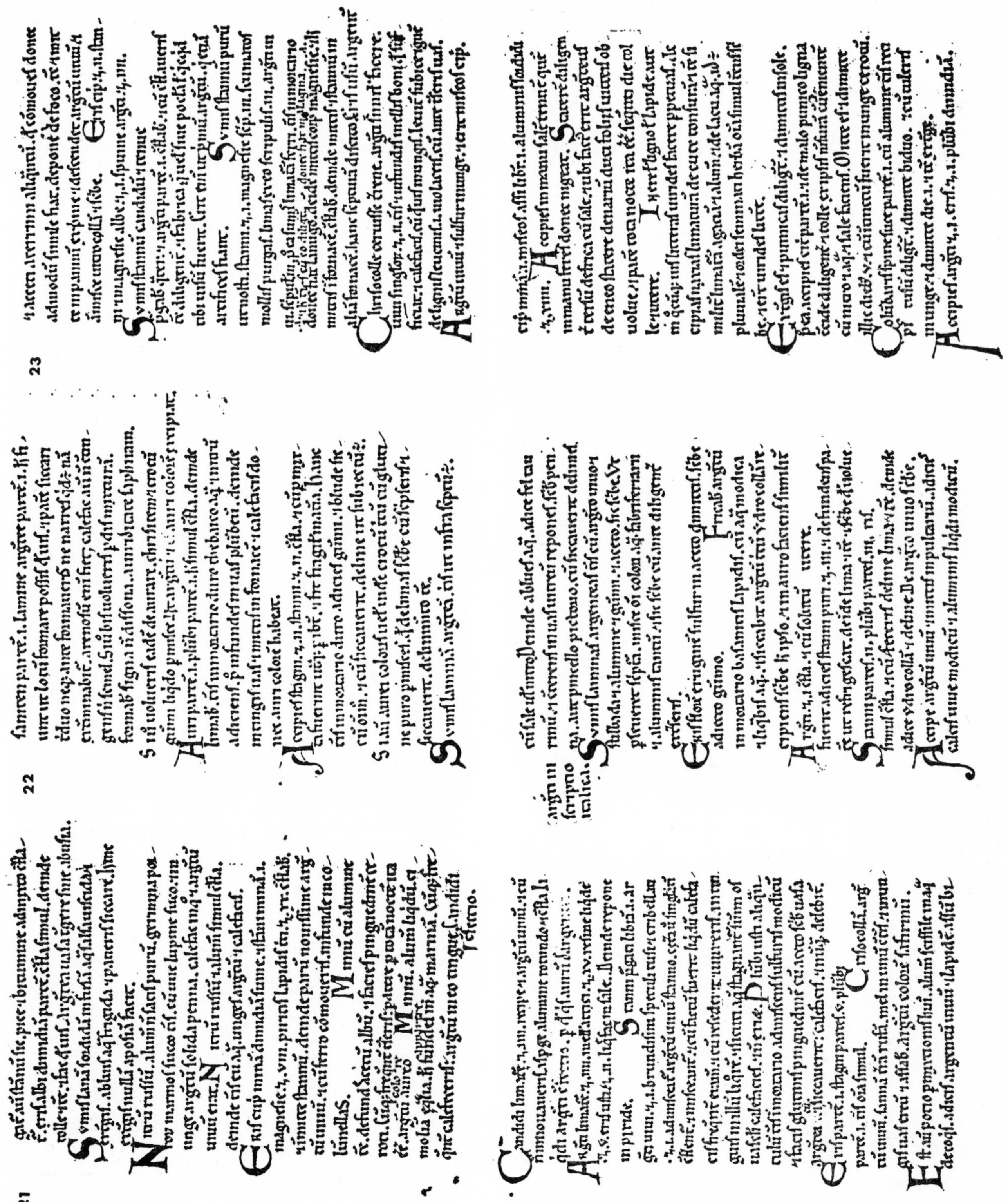

21
22
23

24

cta simul. cũ aũ ungere uel appone gluten et ⁊
calefactũ ferrũ adiunge glutinũ q̄ ẽ appositũ ẽt.
⁊ sic cenab. Sin aũ laminã lamine cõiungis, cere.
Cuprũ p̃ductũ in laminas line colorẽ viridẽ fa
cũ melle ⁊ spuma melli cocta. ⁊ suppone spaculas
ligneas in uase q̃b3 suffundas lotiũ hoĩs. ⁊ stet p̃ dies
.viii. coop̃tũ. Auricalcũ p̃ductũ line
cũ melle cocto ⁊ infunde ei urinã ⁊ acetũ pari mẽsu-
ra. ⁊ sit op̃tũ dies .viii. ⁊ erit q̃si lazur. Indicũ colorẽ
Sucũ de bacis ebuli collige ⁊ diligent̃ sicca facere.
ad solẽ. de hiis q̃ remanserit fac pastillos cũ paruo
aceti ⁊ uini nutrire. Collã grecã facere.
De uernice fac farinã tritã in marmore ⁊ cribratã,
⁊ mitte in ollã rudẽ strictã dilig̃t̃ op̃clo clausã.
ita ut in medio op̃clo sit paruũ foramẽ. ⁊ in ipso
foramine stilus ferreus. ⁊ pone in laminis ferreis
sup̃ fornacẽ aurificis. q̃ p̃us debet incendi. deinde
suppone ligna arida minutissime cõcisa. ⁊ statim
ut incaluerit liq̃scet. Extrahe stilũ ferreũ. ⁊ guttulã
pone sup̃ unguẽ. ⁊ si liq̃da apparuerit subtrahe fo-
cũ. ⁊ infunde olei de semine lini ex p̃ ipsi .ii. par-

tes ad .i. partẽ uernicis. ⁊ iterũ suppositis lignis
coq̃ ad horã paruulã ⁊ utere. Si uõ sit de g̃no mastic̃
liq̃sc̃ tardi9. Inauratio lapid' ł ligñ. Qui aũ uitrũ
inaurat collat partẽ de glutine piscis. ⁊ partẽ de
gũma amigdale. ⁊ miscens coq̃ ⁊ ungue ipsũ uas.
⁊ ĉcide subtilit̃ petalũ auri. pone s̃c̃dm similitudi-
nẽ q̃ uis facere. Similit̃ ⁊ lapidẽ lauans aq̃ ⁊ lignũ.
⁊ dũ siccauerit glutinatio cũ ematite lapide
aut cũ ferro defrica. Ad colorandũ aurũ.
Sume auricũ ⁊ assa ut scis ⁊ aliud tantũdẽ sal
⁊ distempa cũ uino rubeo in eneo uase ĩ numisã
rũ. ⁊ inde line aurũ. ⁊ pone in fornace. ⁊ tãdiu
calefac ut nigrũ fiat. ⁊ extrahe. De dispositione
Dispositio fabrice de pontibus ⁊ q̃b3 fabrice.
mensuris oporteat edificia disponere. ⁊ q̃b3 mensuris in altitudinẽ eleuare. s̃c̃dm modũ fabrice.
Si in altitudinẽ .iiii. staturis fuerit. fabricã unius
stature altitudine oportet ẽẽ. fundamentũ. Si uõ trib3
staturis altitudo usq3 ad bifurcũ erit fundamentũ.
Si aũ unius stature altitudo: usq3 ad geniculũ fundamentũ. Si statim .iiii. cubitorũ. siue .iii. usq3 ad di-

25

di. si duorũ usq3 ad furcã. Si in lignis opertũ in altitudine fuerit: Si uõ talẽ fuerit. q̃ntũ in altitudinẽ.
tantũ ⁊ in fundamentũ debes cauare. ita uidelicet altitudinẽ mensurari oportet. ut tantũ parietis absq3
camera mensuret. Si aũ durus fuerit locus ⁊ montuosus.
cubito uno p̃ staturã pones fundamentũ. Si mollis locus
fuerit: sic sup̃ dictũ edifices. Si uõ petrosus fuerit locus:
ñ crede p̃ eis. s3 caua sic oportet. ne pondẽ nimio
depmat ⁊ subsidat fabrica. De fabrica in aq̃.
Si fabricã in aq̃ necesse fuerit erigere. facies arcã
triangulã. ⁊ picas eã foris cũ sepo ⁊ pice. ut ñ in eã
intret aq̃ ⁊ soluat ipsã calcẽ. ⁊ eos q̃ laborant intus.
⁊ positã arcã int̃ .iii. naues cõstitues in loco ubi necesse fuerit. ⁊ oneratas ipsas naues ut ñ moueant̃ in
aq̃. ⁊ tũ ĩpones lapides ad fabricandũ. Tẽperatio aũ
calcis talis fiat. mittis arene partẽ .i. ⁊ calcis .ii. ⁊ tẽpa-
ris. Ipsa aũ arca habeat .i. cubitũ sup̃ aq̃. De multa.
Multa q̃3 debet ita effici. Mittis calcis partẽ .i. arene
partes .iii. ł .iiii. teste tuse tritã. pulueris palearũ ter-
tiã partẽ. aq̃ uõ ẽgiũ .i. olei porcini sextaria .ii. ⁊ re-
q̃escat ebdomada .i. Si plus dimiseris: melior fiet.

Assidue aũ infundat̃ s̃c̃dm mensurã q̃ indiget. ⁊ effici-
at̃. ⁊ tũ operare. De li camonia. Lacamonia. alumĩ egip-
tiaci. fecis solis .iii. nitri ÷ .i. De cõpositiõe cinnaba-
Tolles ÷ d(r)o argenti uiui mundi partes .ii. sulfuris uiui partẽ .i. ⁊ mitte in ampullã sine fumo ⁊ lento igni. decoq̃ns
facies cinnabarin. ⁊ lauauit̃. De cõpositione iarin.
Tolles petalũ mundissimũ de eramine. ⁊ suspende su-
p̃ acetũ acerrimũ. pone ad solẽ immobilit̃ p̃ .viii.
dies. apies ⁊ tolles ipsũ petalũ. colliges florẽ. facies iarin mundissimũ. De cõpositione psimithii. Tolles plũbũ ⁊ facies petalũ. suspende sup̃ acetũ colliges ipsũ florẽ.
⁊ lauas bene. donec mundus fiat. ⁊ facies psimithin.
Post hec tolles de cinnabarin partẽ .i. de iarin partẽ dimidiã. ⁊ de psimithin partẽ dimidiã. ⁊ mittis in mor-
tariũ marmoreũ. ⁊ teris bene. Post tritionẽ aũ mittis ex aq̃
ubi coq̃t̃ iccio collon. ⁊ fiet pigmentũ pandiũ. Cõpositio
Folia floris uiole colliges. ⁊ in mortario mundo lazurin
teres bene. ⁊ mitte saponẽ ex axungia sine calce mun-
dũ facies. Spuma cũ aq̃ tepida. ⁊ solue saponis in libra
aq̃ ÷ .i. ⁊ fricas subtilit̃ ipsũ saponẽ cũ aq̃ ⁊ dimittis
refrigerari. ⁊ post hec mittis ipsã cõmixtionẽ in tritura

26

florũ. ⁊ mittis illã in uas uitreũ in quo possis mittere ⁊ repones ibi. Alio p̃ aliq̃d tẽpus cõmisces. ⁊ p̃moue illud coti-
die in die semel usq3 ad unã ebdomadã. Post hec .iiii.
dimittes ⁊ duos p̃moues. donec decoq̃s. Post ea tolles
liliũ fuscũ maius q̃d ẽ porfiri. q̃d habet folia ueluti
cultellus. similit̃ ⁊ ipsũ defricas in mortario utilit̃. ⁊ dimittes sine sapone. mittens aq̃. Alio p̃ ea ex uiole cõpositionẽ
libras .ii. de lilio maiore fusco librã .i. aluminis egip-
tii spumati si forte ÷ in spumatũ. saponẽ ut si de-
bile est ⁊ crudũ mittes ÷ .ii. Urine spumate libras
.ii. ⁊ lento igni decoq̃s p̃ horas .vi. ⁊ si multũ uiride
÷ mittes urinã. si uõ multũ uenetũ: plus aluminis mittas. Si aũ lazurin uis facere: ex albo lilio domestico
mittas q̃d sufficit. ⁊ decoq̃s. Gustũ aũ coctionis in ligno unde mouet̃ cõsidera. Calidũ enĩ ñ demonstrat. fri-
gidũ uõ ostendit colorẽ. Coctio aũ lento igni debet
ẽẽ. Agẽtũ cũ coq̃t̃ expendet. q̃ ⁊ ipsa cõmixta debet ẽẽ cũ sapone. s̃c̃dm cõmixtionẽ q̃ p̃dixim9 ⁊ ita
coctioni addi. Quando aũ de fornace toll̃. cacabũ tes-
tũ bitumĩnatũ. aq̃ uult superare cõmixtionẽ.
Aluminis uõ in forti sapone expumato fit. Mitte alu-

mĩ in tepida. dimitte resideat aluminis uisũ. ⁊ effun-
de ipsã aq̃ tepidã. ⁊ ita expuma aluminis. Ipsã aũ
cõmixtionẽ sic defrica. ut in puluere redigat̃. ⁊
ñ sint sp̃es in lazurin. Nã p̃ coctionẽ ipsã cõmix-
tio terenda ÷ ⁊ mouenda. ⁊ in umbra desiccanda
in ipso cacabo. ⁊ post ea ad solẽ eicienda. ⁊ torrenda
ut fiat lazurin. ut picta aq̃ deleri ñ possit.
Oleo q̃d appellat̃ cicinũ sup̃ pictũ ad solẽ pinge. ⁊ ita
cõstringit̃. ut nũq̃ deleri possit. Cõfectio pandiũ.
Folia flor' papaueris mg̃ collecta repones in pinna-
to nouo. ⁊ operiens pones ad solẽ .i. die. ⁊ dũ marcei-
da fuerint tolles aq̃ in q̃ iccio colla decoq̃t̃. ⁊ mitte in ipsã folia florũ ⁊ teres utilit̃ miscens cũ modico
cinnabarin. ⁊ erit color pandius. Cõfectio ficarin.
Tolles laccã mundã ⁊ teres mundit̃ ⁊ coq̃s eã in urina spumata. ⁊ iottã q̃ exierit refundes in uas mũdũ
uitreũ. ⁊ post ea tolles siccũ ⁊ bene tritũ puluerẽ de lilio albo domestico. Si aũ infuscauerit ñ mittas nefus
cũ fiat. istud tritũ in libr. .ii. de unico flore ⁊ ipsũ
tritũ libr. .iii. pulueris duarũ cõmixtionũ unicuiq3
coctioni lacce ÷ .i. de alumine egiptiaco cõmis-

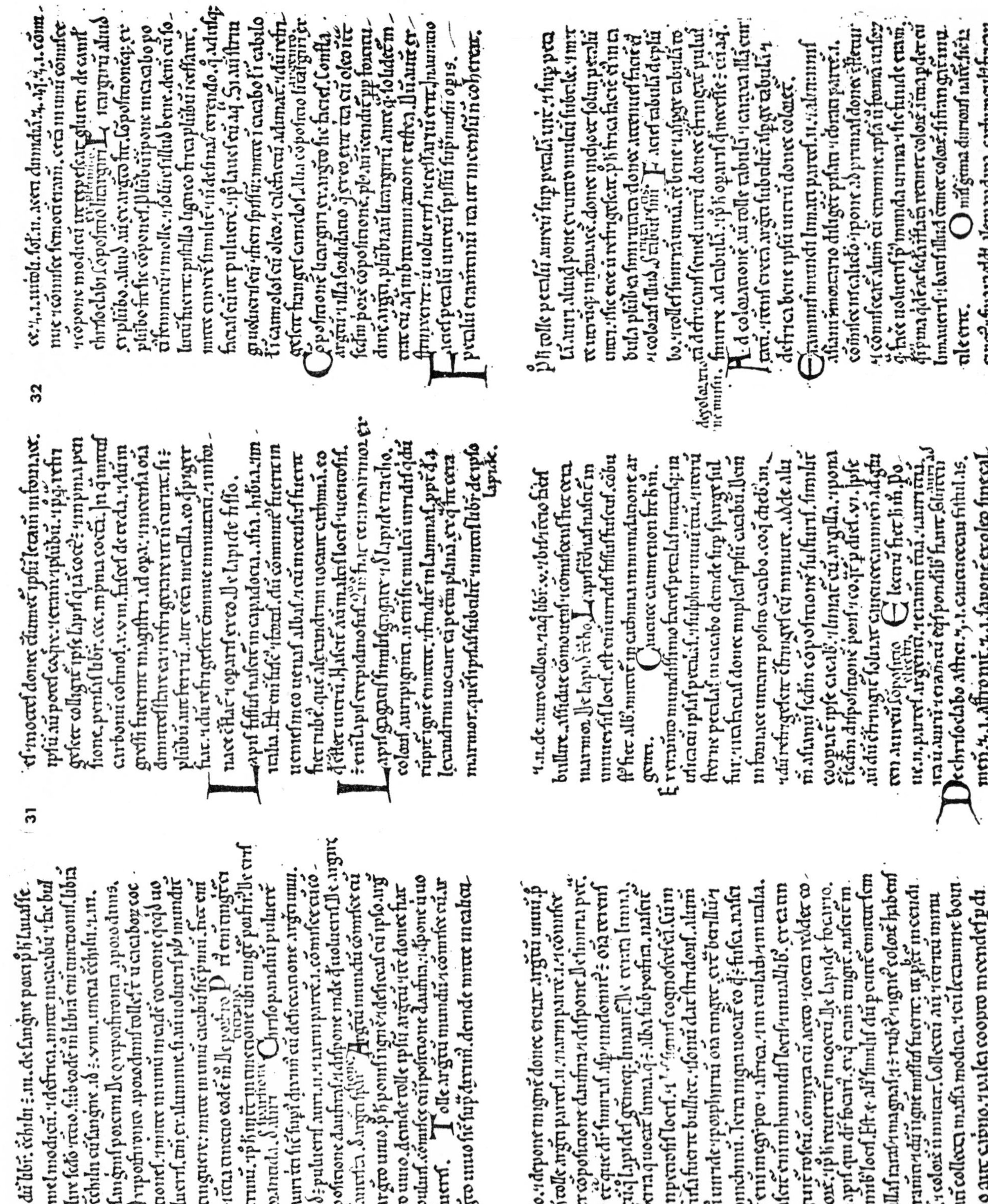

36 coque quo adusque consumetur urina et pinguescat ipsa coctio. et mitte refrigerare. deinde sume
de floribus neulacis libr. iii. de cocto autem guatto
libr. ii. de cinnabarin dimidiam z. et commiscens
tere minutim. et trita coopta in mortario stare per
noctem. deinde ostreas diligenter mundatas intus
et foris et lotas a sordibus et limo ponas in cacabo novo et decoque usque ad pulverem. et refrigeratas tere
diligenter seorsum tollis ex ipso pulvere libr. i.
et larin mundum mitte in aliam urinam dispumatam
terens diutius donec turbetur et virideseat urina. et
ex eo commisce in proprio mortario cum supradictis speciebus. et bene defricans repone in vase novo ad solem una die. postea coopertum vas oblimes diligenter.
et pones in fornacem vitrarii superiorem die i. et exiet
lazurin. Flores neulacis cum sapo-
ne fricatos ut supradictum est infunde in urina spumata. et vase cooperto pone in letamen ut fragidet.
Similiter folia guatti infunde in urina spuma- r
ta et reposita in letamine. cum fragidaverit. eice de
ipsis in mortario auferens ex ipsis omnia nervia.

et tolles ex ipso guatto libr. i. et de flore neulacis
libr. ii. et papaveris z. ii. et commixta tere. addens
cinnabarin ÷ dimidiam. et larin ÷. dimidiam. urine
expumate. ÷ dimidiam. tere diligenter. et mittes
in cacabum novum. lento igni decoque donec consumetur
et pinguescat et fiet lazurin modicum porfirizonta.
Flores neulacis infunde aceto
et compositos in vase cooperto ut predictum est pone in letamen ut fragidentur. Similiter guattum exnervatum tere diligenter et in vase novo aceto infusum operies in letamine donec fragident
ipsa folia. postea sume de guatto libram i. de
lacca decocta in urina spumata ÷. i. et ex flore neulacis libr. i. de pulvere ostree munde
et lote sicut supradictum ÷ z. iiii. his omnibus in mortari-
o tritis. de coctione herbe celidonie quam in
urinam coxeris libr. dimidiam. adde et croci ÷. i.
Omnia trita et in uno vase coopta et in letamine
die i. posita tolles. et ad solem siccabis et uteris. lazurin acrum. Tolle florem de neulace et defrica
cum sapone sicut supra documentum est coopertum in vase re-

37 pone in letamine. Similiter facies de guatto enervato. et post dies cum putruerint. tolle
de neulace et guatto z. i. et terens subtiliter
in mortario addens spumithin mundi ÷. i.
et lulacin confecti et mundi ÷ 5. cinnaberin ÷ 5.
et urine expumate cum vitolo trito ÷. iiii. urine autem libr. x. et postquam residerit vitolum tere ipsi
urinam in mortario quantum sufficit et commixtum
dimitte residere ii. dies. post hec tolles urine
expumate munde libr. iii. galle tere z. i. commiscens dimitte infundere die i. deinde mitte
ex ipsa tota libr. i. et ita bene dimitte residere ad solem. et fiet lazurin acrum. Item aliud.
Tolle lazurin primum ÷. i. cinnabarin ÷. i.
compone ut supra lazurin vel colont. Pspumithin
mundi tere libr. i. lazurin z. i. cinnabarin.
÷. i. compone ut supra lazurin vel limzonta. Sume neulacis flores unctos ex sapone ut supra monstratum est
÷ libr. i. et pone in letamen. et guattum exnervatum
cum sapone ut supra composuisti. atque in letamine maceratam libr. i. postea tritis in mortario ad-

de cinnabarin z. i. herbe lucie ex cocte que co-
quenda est cum urina expumata. donec veniat
ad iii. partem. et donec pinguescat ex ipsa pin-
guedine trita libr. i. hec omnia commixta pone ad solem et fiet lazurin melinizonta. Alia lazurin.
Cinnaberin ÷. i. siricum ÷. i. lacce coctionis ÷. i. lacca autem coquitur sic. laccam subtiliter tritam coques in urina
expumata bene lento igne. et ex ipsa coctione
sume libr. i. lulacin sol. i. et trita simul dimitte residere et siccari ad solem. Item. Cinnabarin
÷. i. lulacin sol. ii. psimithin sol. i. ita bene sicca ad solem compositum. Ex suco florum papaveris ex-
pressi ÷. i. cinnaberin vel vermeli ÷ 5. lulacin sol. i. hec omnia
commixta et trita sicca ad solem.
Cinnaberin vermiclin libr. iiii. vermiculi terrenum qui in foliis cerri nascitur libr. i. de coctione
lacce supradicte libr. i. urine expumate. lib. x.
et sumptum utrumque vermiclin subtiliter tritum. mitte in cacabum in ipsa urina in lincheolo raro. delava vermiclin in cacabum in quo urina decocta est.
et iterum tere coctum et delava in urina cacabi et

38 sic facies donec consumetur coccus totus. deinde
coques diligenter commixtionem illam et exagitas. Item.
sume ostream mundam et bene lotam. et mittes eam in
spumatam bene coopertam. ponasque in furnum donec
refrigeretur. et postea tere subtiliter. atque ex ipso pulvere
libr. iii. mittas in coctionem predictam bulliat bene usque in tertio. Deinde repone ad solem ut
pinguescat. Pandius. Mitte vermiclin libram
i. coccarin libr. i. coccarin nascitur sicut supra dictum
est in foliis cerri. cinnaberin ÷. i. lazurin primi
÷. i. commisces tere diligenter in mortario et
mitte de urina expumata libr. xv. coques in
cacabo novo donec ad dimidiam partem veniat
ipsa urina. postea pisa grana cum cinnabarin trita
in lincheolo delava sicut supra continentur donec
consumetur. Item pandi. Vermiculi libr. 5. de alio
vermiclo ÷. vi. psimithin ÷. vi. lazurin ÷. vi.
hec diligenter trita mitte in cacabum cum urina
dispumata libr. x. et mittens in lincheolo raro coccum delava in urina. et iterum delava donec expendatur coccus. et decoques donec tem-

at urina ad dimidiam partem. et repones ad
solem. Lulacin libr. i. cinnabarin. libr. i.
psimithin. libr. i. ficarin ÷. ii. hec omnia trita et
cum aqua tepida mixta pones ad solem donec siccentur. Lulacin libr. i. cinnabarin principal
÷. i. lazurin. libr. i. ocrea mundissima libr. q
anus libr. i. hec omnia bene trita et cum aqua tepida
mixta defricans pones ad solem donec siccentur.
Lulacin ÷. iii. psimithin ÷. iv. Lulacin
libr. i. quiam libr. i. psimithin ÷. i. Lulacin
÷. i. ficarin ÷. i. quiam ÷. i. Quiam ÷. ii.
psimithin ÷ dimidiam. commixta tere cum urina
despumata quod sufficit. pone ad solem.
Quiam ÷. i. ficarin ÷. i. ocree ÷. ii. omnia trita commisces cum urina dispumata et repones ad solem.
Quiam ÷. i. de pulvere cauicuceetamenti triti.
÷. i. ficarin ÷. i. ocree ÷. i. omnia trita et commixta
cum urina dispumata pones ad solem.
Cinnabarin libr. i. herbe lucie coctionis ÷. i. croci clari mali ÷. i. ficarin ÷. ii. quiam ÷. i. omnia
trita commisces urine expumate et sicca ad

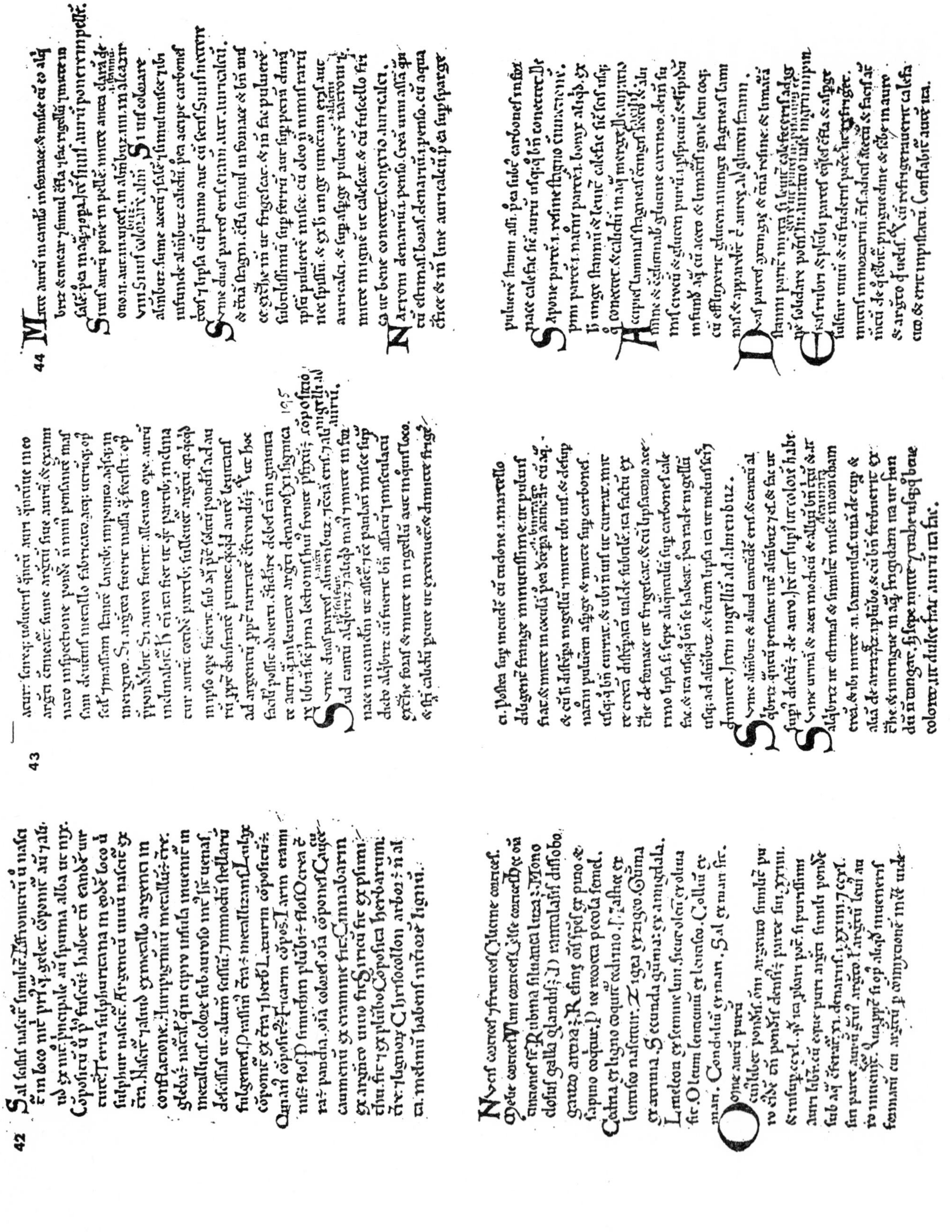

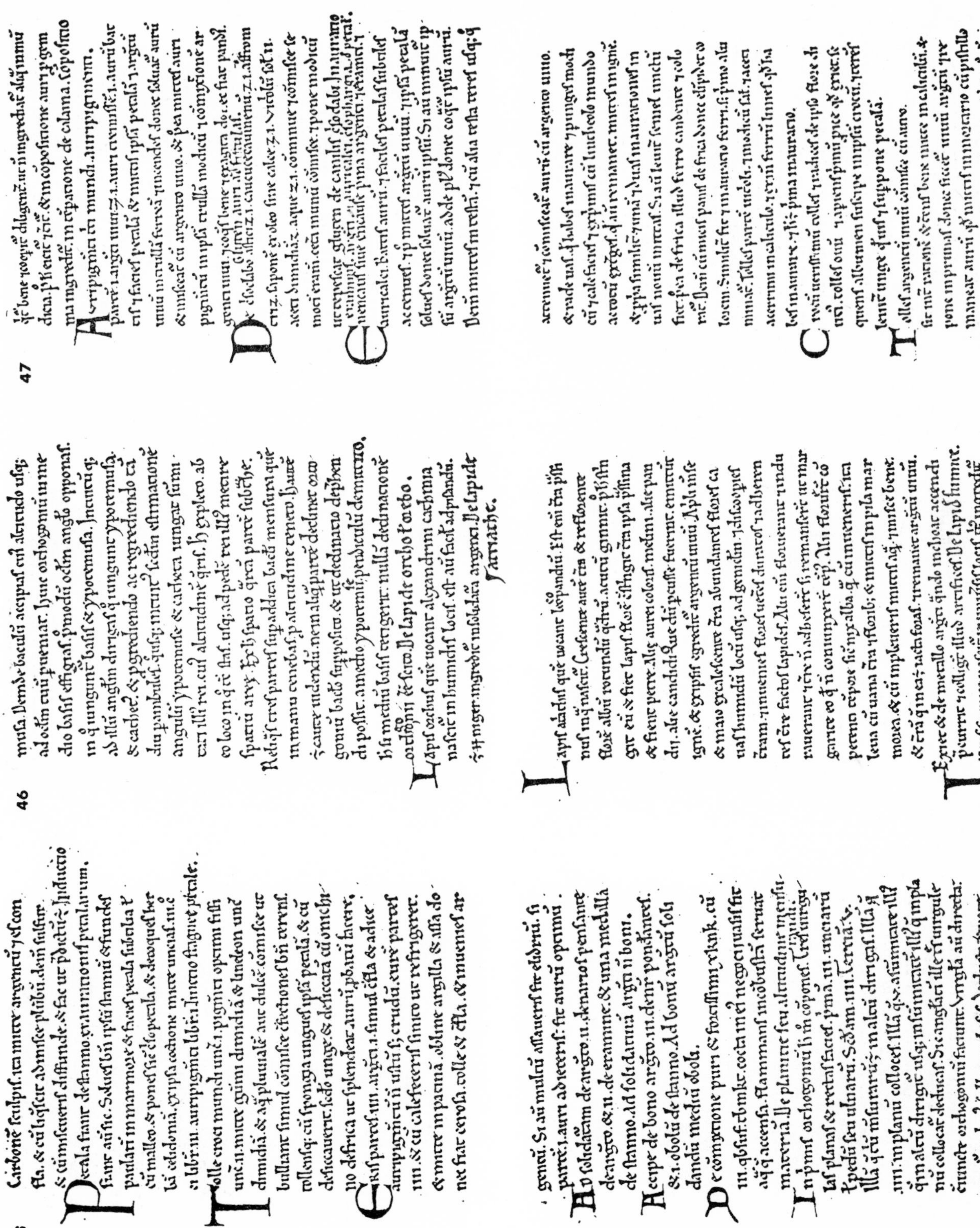

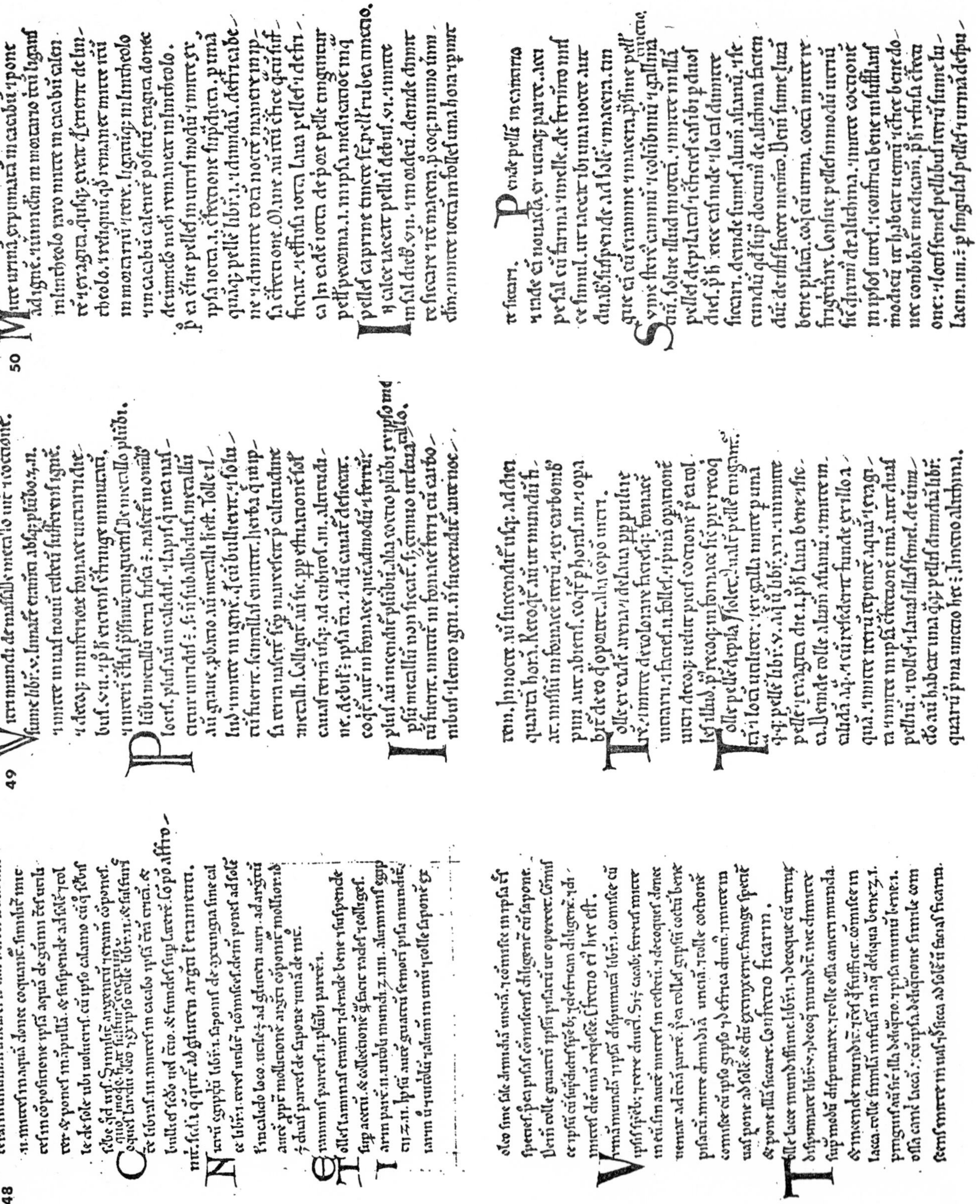

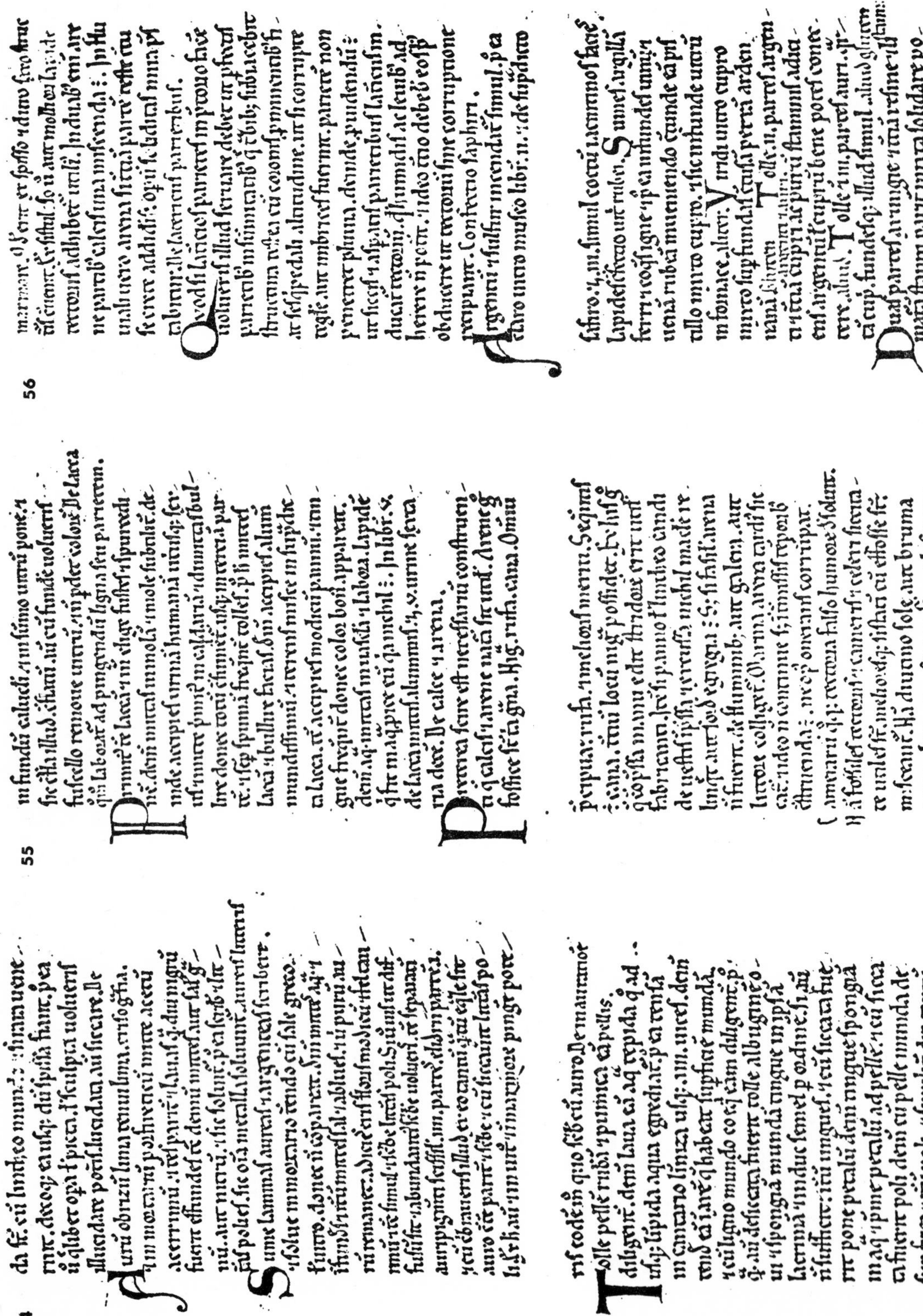
54
55
56

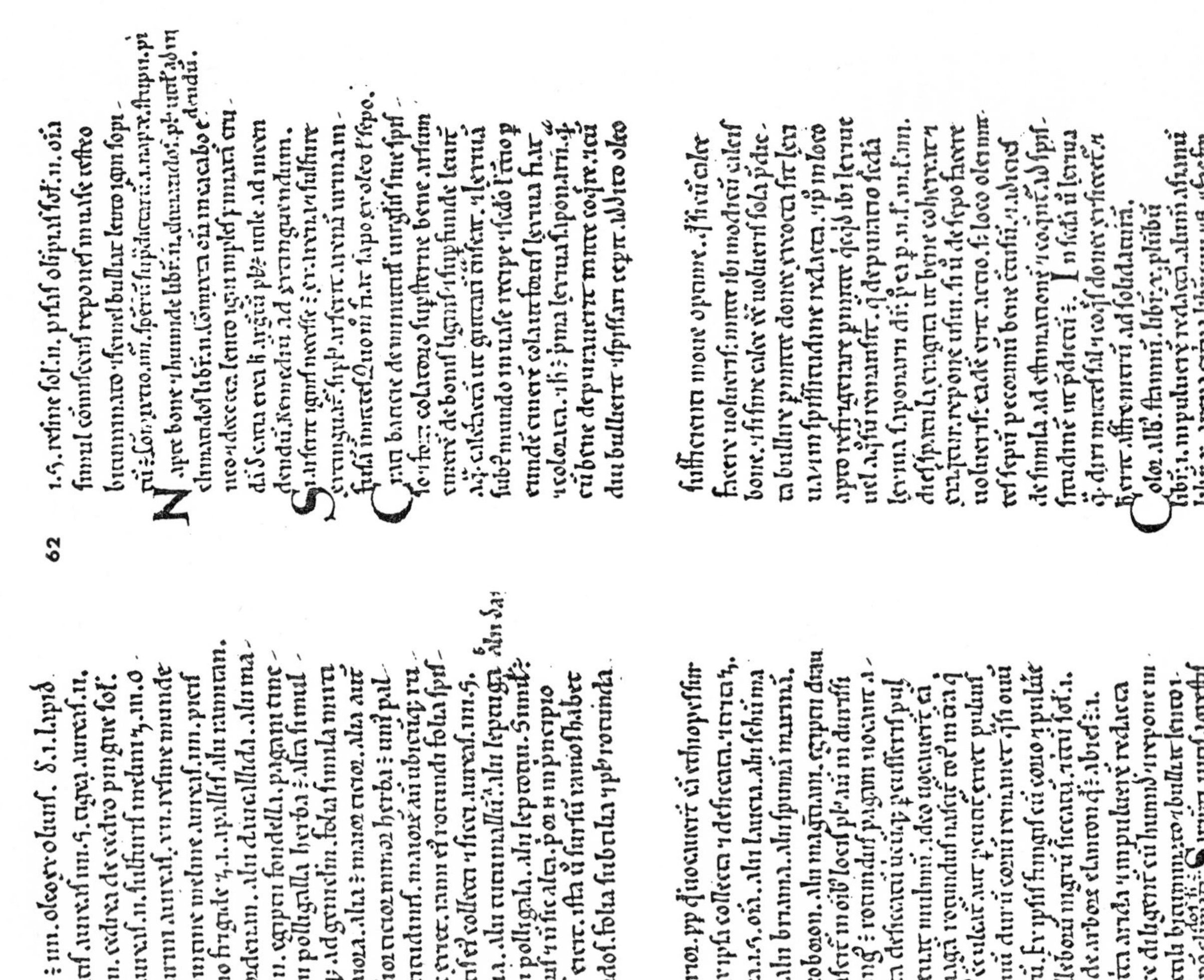

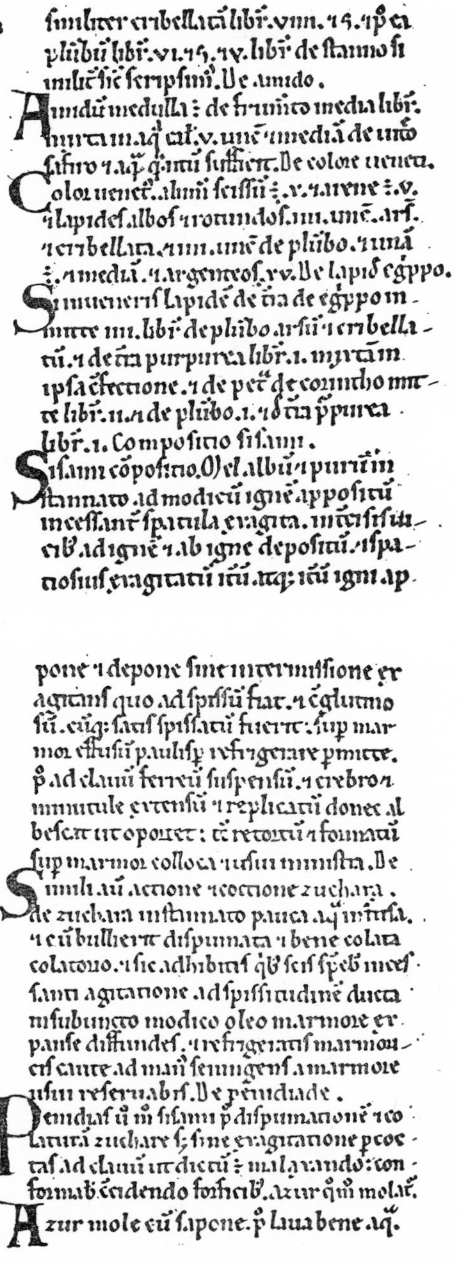

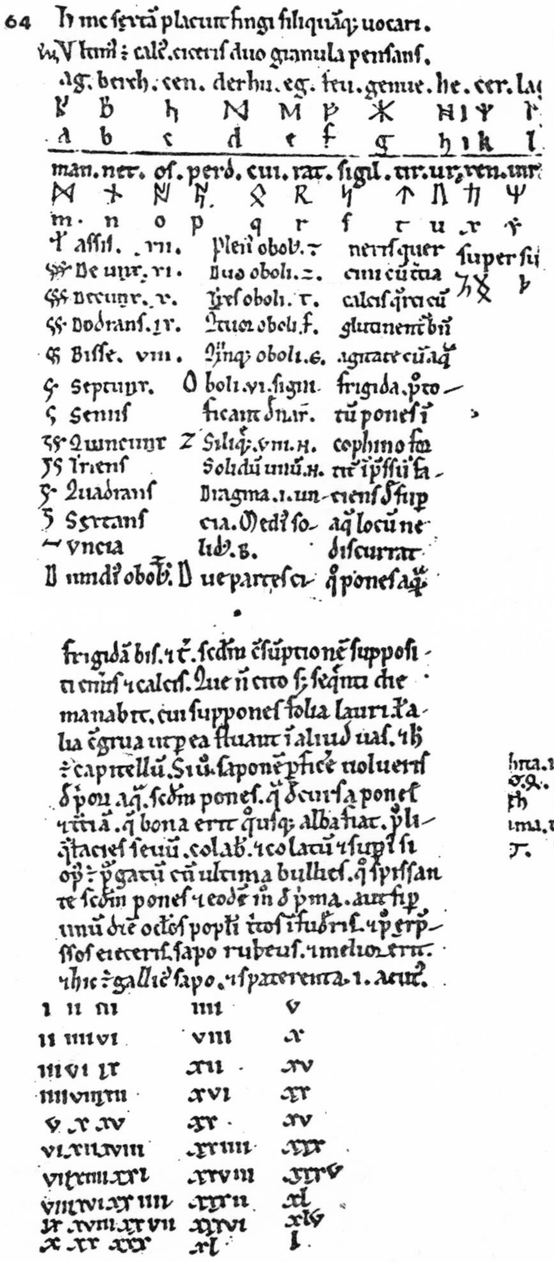

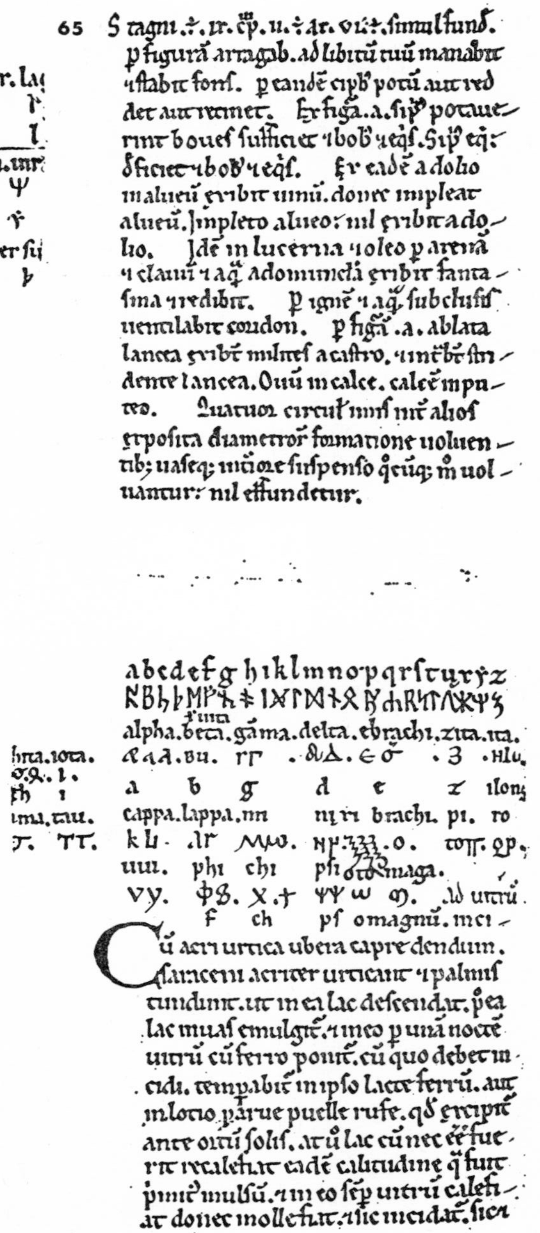

APPENDIX C

A NOTE ON THE RUNES

Eric P. Hamp

Department of Linguistics, University of Chicago

The runes in chapters 288-B and 288-P are clearly in the Anglian tradition and are of considerable importance in runic studies.

The most recent full treatment of English manuscript runes is R. Derolez, *Runica Manuscripta: The English Tradition* (Brugge, 1954), wherein he discusses (219–237, esp. 227ff) Phillipps 3715 and the related Exeter 3507 (tenth century), from which surely is derived Cotton Vitellius A 12 (eleventh–twelfth centuries). Derolez did not know the whereabouts of Phillipps and worked from the printed version in *Archaeologia*. Though my judgment is independent and there are a few points on which we differ, Derolez and I agree well on the whole.

In writing the following notes, I have also depended heavily upon H. Arntz, *Handbuch der Runenkunde* (2nd ed., Halle/Saale, 1944) which discusses English runes especially on pages 120ff, 173ff, and rune names on 188ff. On runes generally see Klaus Düwel, *Runenkunde* (Stuttgart, 1968) (strongly influenced by Krause), who gives an up-to-date bibliography and some slight references to our class of documents on pp. 104ff.

Despite the great difference between the two alphabets given in the *Mappae Clavicula* (chaps. 288-B and 288-P), they are mutually clarifying and will be considered together, thereby bringing out much of the tradition that is needed to grasp the correct focus of the separate signs and names. It seems clear that the origin of the first alphabet (chapter 288-B) is an Anglian version of a *fuþorc* which has passed through the hands of High German speakers on the Continent, and was arranged into an alphabet by someone not very directly learned in the tradition of runes. I see this alphabet as closer in relationship to Munich 19410 than Derolez seems to. Thus, I see a *ʒār* rune as underlying the Phillipps alphabet, reminiscent of the use of *ʒār* in the role of *z* (by continental misreading of *ʒ*) in Vienna 751 (see Derolez 205).

With such a transmission it is hard to say anything crisp or penetrating about chronologies; there are too many strands of unequal certainty and documentation, and at least one scribe of shaky learning lies in between. Nevertheless, the relation to Ex 3507 makes it likely that we have here a continental alphabet which had been brought back to England, though of course originating in English learning in the first place. Since in my view the alphabet pre-supposes *ior*, *calc*, and *ʒār* we are dealing with a late version of the *fuþorc*. This would be consistent with a tenth-century immediate *Vorlage*. If there is a relation with Munich 19410 and also with Vienna 751 (both ninth century), then the directly pertinent continental antecedents go back a good century or more earlier.

Some detailed comments on the alphabet in chapter 288-B follow:

a *ag* is clearly *āc* f. 'oak'. Other instances with *g* are known. The writing of *g* for *c* would be easily explained if, as other evidence suggests, a tradition is being drawn on that went through High German hands. That is, we would have a back-formation just as is found with *berg* (reinterpreted as 'mons') for *berc* 'birch'. The basis of this is the High German consonant shift whereby for a *c* a German expects Old English *g*, thus creating hyperforms. This possibility is enhanced in this word since the diphthong seen in German *Eiche* had been monophthongized prehistorically in OE *āc* (just as in *Stein: stān* 'stone', etc.), thus rendering the word scarcely recognizable to a German speaker.

b *berch* is clearly *beorc* f. 'birch', but it is worth noting that all forms in this alphabet show "unbroken" English vowels (*cf. eg, perd*). If we are to take this consistent feature at face value it places the present alphabet in the tradition of the Anglian transmission.

c *cen* must be *cēn* m. 'torch'.

d *derhu*. The expected name for *d* is *dæʒ* and the rune is an unmistakable *d*. If however we note that *þ*, 'thorn', appears in the unnamed runes at the end of the table, a solution immediately suggests itself. The name *thorn* was an innovation of the English runes; the old Germanic name was **þuris* 'giant' and the like, depending on dialect. The latter is found, for example, in the Leiden Codex (tenth century, from about 830) written as *dhurs*. Moreover, a German would render *þorn* m. as *dorn* in his own language, with the familiar change of voiceless *th* to *d*. Further, *u* is an easy misreading for *n*. Thus a conflation of **dhur(is)* with **dorn* or **dhorn*, plus a familiar misreading, could easily yield *derhu*. If so, we have here in concealed form the name of *thorn*.

This hypothesis seems to be confirmed when we note that the alphabet in chapter 288-P actually shows *þ* but miscalls it *d*. It looks like an error made by a Continental West German.

e *eg* must be *eh*, that is *eoh* m. 'horse'. The "unbroken" *e* has already been noted, but in this case, as previous scholars have remarked, the form *eh* was generally accepted to avoid homophonous clash with *ēoh* 'yew' (rune No. 13). The reason for *g* in place of *h* is not clear, but if it does not represent the runic conflation seen in alphabet 288-P between the signs for *g* and *h*, then it may reply to the fact that *h* no longer was heard medially in English words but originally alternated grammatically with *g*.

f *feu* is clearly *feoh* n. '(movable) property.' Although the spelling here points to the Norse tradition (matching that of the *Abecedarium Nordmannicum* transmitted early in the ninth century by Hrabanus Maurus), Derolez thinks *feu* is more likely Old Low German (ultimately Old High German) rather than Norse, since no other Norse traces are clearly present.

g *genue* must be *gyfu geofu* f. 'gift,' but the details are not clear. We have an attested OE orthography *ʒywu* and Germanized variant writings of *gebo*. If we hypothesize a variant with the Germanized or "unbroken" vocalism *e*, but with an orthography like the Salzburg *geuua*, we may regard the *n* as a misreading for *u*. But the final *e* is hard to be precise about.

The shape of this rune is revealing. It is nothing other than No. 33 *ior* ~ *iar*, which properly represented *īo/īa*. This was a late rune and a borrowing from Scandinavian (see Arntz 122, 232). The rune *g* (No. 7) represented in the English system the palatalized *yod* sound which had resulted from Germanic *g* before front vowels. Thus we see that the *i*-diphthong had been mustered into service as a clarifying device; this may have been a Continental phenomenon, applied by someone to whom the English *ʒ* and diphthongs were not native.

h *he* suggests the correct name *hæʒl* m. 'hail.' It seems that the vowel *e* actually represents the graph *æ*, but the rendition was simply mutilated, in an earlier copy. Note here that the *Abecedarium Nordmannicum* and other sources show *hagal*.

i The lack of the name *īs* m. 'ice' must result from the same mutilation in the exemplar that truncated the previous identification, as Derolez has convincingly shown.

k This letter is extremely interesting. We must remember that Germanic *k* was palatalized, like *g*, before front vowels in English; the result of that value is reflected in *c* (*cēn*, No. 6). I therefore interpret Υ as an inverted ⋏ *ᵢcalc* (No. 31), or so-called *k*[II]. This letter was intended to represent the distinctive unpalatalized *k* in late Old English.

The name *cer* seems to reflect a garble and conflation of the name *ʒēr*, as we find No. 12 *ʒēar* n. 'year' in several sources, the *ʒār* m. 'spear' (the late rune No. 33 for unpalatalized *g*[II]) and the alternation of *g* and *c* already noted under *a*. In other words, *calc* has been renamed on the model of *ʒār*, while *yod* and *g* have been confused as we have seen with *ior* above. I do not accept Derolez's suggestion that this is an *x*, in view of the added testimony of chapter 288-P and of the need to account plausibly and fully for the letter shape.

l *lag*[*u*] m. 'lake' is exactly what we expect, although Derolez also allows the possibility of lago.

m The sign should have been written with long legs, ᛗ. It is now the shape properly of *d*. The same error is made in chapter 288-P.

n *net* must be for *nīed* f. 'necessity.' The exact accounting of the vowel is not clear, but the *t* looks like a Germanism. This may be viewed as comparable to the German vocalism of *nōd* which is frequent in English futhorks. This and other names seem to me to suggest a very close relation to Munich MS 19410 (ninth century), which also shows *ag*, *geuo* (underlying *geuue?*), *heih* (crossed with *is* and representing the truncation seen above?), *ker*, *n&*, *perd*, *rat*, *uyr*. In this connexion, Munich 19410 also has *caar*, deriving somehow ultimately from *ʒār*, a further support for my assumption of *ʒār* above.

o *os* is surely *ōs*, which came to be regarded as Latin *ōs* 'mouth.' But in origin, of course, the rune was Germanic *a*, and the name was the divinity **ansuz*. *āc* and *æsc* 'ash' (not in our list) are later offshoots of this.

p *perd* is for *peorð*, for which we have attested variants *peord* and *perd* (St. Gall). There is much debate on the origin (Keltic?) and meaning of this rune name, but the present instance offers no problem.

q *cui*: The very presence of a rune with such a value, apart from the dubious name, is simply a schematization on the Latin alphabet. Germanic had no *q*. There actually is a sign Υ attested, with the name *cweorð*, but this must be a late invention modelled on Latin; see Arntz 124, 209. The fact that *cweorð* was patterned on *peorð* makes one additionally wonder whether behind this there might have been some Irish monastic learning that knew the Keltic equivalence between early Irish *q* and the British Keltic (e.g. Welsh) *p*. Derolez points out that *cui* occurs elsewhere as *cur*. This then fits with *cer* and *ʒār* on an acrophonic principle.

The sign ᛜ is of course *o* or *œ*, No. 23 *ēþel* (ca. 800), *œþil* (ca. 600) < **ōþila-* 'property.' Perhaps the shape simply suggested Latin Q.

r *rat* is certainly *rād* f. 'ride.' The form with *t* strongly suggests the *Abecedarium Nordmannicum rat*, with High German *t* for *d*. Moreover, after having been Germanized in form, the meaning of this name was sometimes later taken as being 'counsel' instead of 'ride.'

s *sigil* is the expected late unrounded development of *sygil* n. 'sun.' This is linguistically a very archaic word.

t *tir* is the presumed late unrounded development of *Tȳr* (the divine name fossilized in *Tues-day*), which is supposed to show Norse borrowing with the −*r*, but there is a complex debate and uncertainty surrounding the exact background. (See Arntz, 216–219).

u *ur* is *ūr* m. 'aurochs,' the European bison just as we should expect.

x Both the sign and name pose problems, but not, I think, insoluble if considered together. The sign appears be to me to an inverted variant ⅃ which is attested (see Arntz 123) for *s*, the more common shape of which is ⅃. Now it is interesting and, I think, pertinent to note that in the late Dal-runes (Dalekarl runes; see Arntz 113), where many letters have been overtaken by Latin shapes, both *c* and *x* share the same shape ⅃. Thus *x* seems on the one hand to be treated as a variety (in Latin pronunciation?) of *s*, and on the other as an orthographic relative of *c*.

Having observed that *x* and *c* are closely related, we are not surprised then to find the name *xen* (however it was supposed to be pronounced!) modeled transparently on *cen*. This all must be a very late invention, and points to an English basis embroidered with continental learning.

y The sign is not the expected one. Ψ looks very much like No. 15, which is Old English *x eolhx*, displaced one position, and perhaps also influenced by No. 8 ᛈ, the *wen* sign for *w*, also attested as *uyn* (*wynn* f. 'joy').

The name *uir* is presumably *ȳr* m. 'bow,' but also is reminiscent of Brussels and St. Gall *yur* (pronounced [ür]?). Perhaps these spellings helped to draw in *uyn* (*wynn*). This is all speculative; it is hard to know how many elements are involved in such a conflation.

z This sign is opaque. Is it not perhaps just a different written form for *s*? Certainly *z* was always of dubious status. The missing identification of the last three runes suggests again the mutilation of the copy. A *z* that would have been proper here appears meaninglessly before the entry *Siliquae viij*. H in the table of weights that immediately follows, and was perhaps misplaced by [an earlier?] scribe. It should be noted that the rune *æ* is lacking from this alphabet, perhaps by mutilation; *cf.* Vitellius A12 and Exeter 3507.

ng This is the expected *Ing* rune.

th Here we have the well known *thorn* rune. Both these last remain without name because they did not belong to the Latin alphabet; here is a clear testimony to the break with the old *fuþark* tradition.

Except as noted in the following (and with the omission of *ng*), the alphabet in chapter 288-P agrees basically with the one given in chapter 288-B. Some comments on the letter shapes are, however, in order.

g As Derolez (236) points out, this is a *g* of the *isruna* type (122–123).

h The shape here is noteworthy in departing from that of the preceding alphabet, which is surely more conservative. This seems to be the type of the Danish ⋇ (*ca.* 900–1300), and the parallel cross-bars are strikingly like those found on the Älfdalen stave of *ca.* 1750! Derolez links it with an *ing* rune.

k This is simply *g*ᴵᴸ*-ʒār*. This lends support to our argument on the name of *k* in the note to chapter 288-B. Derolez posits an intermediate *car*.

o This shape is dealt with under *q* in the above note. Its value here is a piece of transparent archaizing: First, the value *o* (as opposed to *ǣ*) is prehistoric; secondly, this shape has ousted from the list the very old letter ᚩ *ōs*, which had prehistorically developed from a value *a* to *o*. Derolez takes the phonetics at face value and argues a Continental transmission.

q This is simply *k*ᴵᴵ *calc*.

x This is simply No. 33 *ior*, which we have already seen above with the value *g*. We see from this how the more usual shape of *h* has been displaced above. See, however, Derolez, p. 236.

The preceding peculiarities, heterogeneous and perverse as they may seem at first, have an internal consistency which is instructive once it is grasped. To recapitualte, *g* may perhaps be *x* (which we have already identified with *c*); *h* shows a shape historically overlapping *g*; *k* is the new unpalatalized *g*; *q* is the new unpalatalized *k*; *x* is a variety of (palatalized) *g*. Thus we have in effect two orderly replacement series of functions affecting exclusively the velars and the palatals:

$$x(=c) \xrightarrow{\text{ᚻ}} g \xrightarrow{\text{⋇}} x \xrightarrow{\text{ᛉ}} h$$

or, more precisely, $x(<c) \rightarrow g \rightarrow h(> \text{ᛉ})$; $g \rightarrow x$ and

$$g^{\text{II}} \xrightarrow{\text{⊠}} k^{\text{II}} \xrightarrow{\text{ᚼ(ᚼ)}} q$$

We may see two factors motivating these replacements: (1) The need to acquire a rune *q* to match the Latin alphabet (which the previous alphabet did by robbing *œ*), and to bolster the chronically weak *x*; (2) The problem of distributing the Old English palatalization (*ċ* and *ʒ̇*, vs. *k* and *g*) with respect to the Latin availability of both *c* and *k*.

BIBLIOGRAPHY

ALEXANDER, SHIRLEY. 1964–1965. "Medieval Recipes Describing the Use of Metals in Manuscripts." *Marsyas* **12**: pp. 34–51.

——. 1965. "Base and Noble Metals in Illumination." *Natural History* **74**, 10: pp. 31–39.

ARNTZ, HELMUT. 1944. *Handbuch der Runenkunde* (2nd ed., Halle).

BAILEY, KENNETH C. 1929, 1932. *The elder Pliny's chapters on chemical subjects* (2 v., London).

BAXTER, JAMES HOUSTON, and C. JOHNSON. 1934. *Medieval Latin Word-list from British and Irish sources* (Oxford).

BERGER, E. 1912. *Beitrage zur Entwicklungs-Geschichte der Maltechnik* (3 v., (Munich).

BERTHELOT, MARCELLIN PIERRE EUGÈNE. 1906. "Adalard de Bath et la Mappae Clavicula. . . ." *Journal des Savants*.

——. 1893. *La chimie au moyen-âge* (3 v., Paris; facsimile reprint Amsterdam, 1967).

——. 1887–1888. *Collection des anciens alchimistes Grecs* (3 v., Introduction, traduction et textes, Paris; facsimile reprint, London, 1963).

——. 1889. *Introduction à l'étude de la chimie des anciens et du moyen-âge* (Paris; facsimile reprint, Paris, 1938).

BIRINGUCCIO, VANNOCCIO. 1540. *De la pirotechnia* (Venice). English translation by C. S. Smith and M. T. Gnudi (New York, 1942; reprinted 1943, 1959, 1963).

BRILL, ROBERT H. 1970. "Chemical Studies of Islamic luster glass." In R. Berger, editor, *Scientific Methods in Medieval Archaeology* (Berkeley & London, 1970), pp. 351–377.

BURNAM, JOHN M. 1920. *A Classical Technology Edited from Codex Lucensis 490* (Boston).

——. 1913. *Recipes from Codex Matritensis A 16* (University of Cincinnati Studies, ser. 2, Vol. 8, part 1).

CALEY, EARLE R., and JOHN C. RICHARDS. 1956. *Theophrastus on stones:* Greek text, English translation and commentary (Columbus, Ohio).

——. 1926. "The Leyden Papyrus X—an English Translation with Brief Notes." *Journal Chem. Education* **3**: pp. 1149–1166.

CARDALUCIUS, JOHANNES H. 1672. See Ercker, Lazarus.

CELLINI, BENVENUTO. 1568. *Due Trattati, uno intorno alle otto principali arti del l'oreficeria. L'altro in materia dell'arte della scultura* (Florence). English translation by C. R. Ashbee (London, 1898).

CENNINI, CENNINO. 1932–1933. *Il libro dell'arte* [1437]. Edited and translated by D. V. Thompson, Jr. (2 v., New Haven).

CLAGETT, MARSHALL. 1959. *The Science of Mechanics in the Middle Ages* (Madison).

CURTZE, M. 1895. "Die Handschrift No. 14836. . . ." *Abhandlungen zur Geschichte der Mathematik*, Heft 7: pp. 138–139.

DARMSTAEDTER, ERNST. 1926. *Berg-, Probier- und Kunstbüchlein* (Munich).

DE VAUX, CARRA. 1903. "Le livre des appareils pneumatiques et des machines hydrauliques de Philo de Byzance. . . ." *Mem. Acad. Inscriptions et Belles Lettres, Notes et Extraits* **38**: pp. 27ff.

DEERR, N. F. 1949–1950. *The History of Sugar* (2 v., London).

DEGERING, H. 1917. (Foreword by Herman Diels). "Ein alkohol rezept aus dem 8. Jahrhundert." *Sitzungsberichte der Königlichen-preussischen Akademie der Wissenschaften* **36**: pp. 503–515.

DEROLEZ, RENÉ. 1954. *Runica manuscripta* (Brugge).

DIELS, HERMAN. 1914. *Antike Technik* (Leipsig).

——. 1913. "Die Entdeckung des Alkohols." *Abhandlungen der Königlichen-preussischen Akademie der Wissenschaften. Philosophisch-historischen klasse*, p. 12.

DIOSCORIDES. 1907–1914. Ed. Max Wellman, *Pedanii Dioscuridis Anazarbei de materia medica libri quinque* (3 v., Berlin).

——. 1934. *The Greek Herbal of Dioscorides . . . Englished by John Goodyear A.D. 1655*, Edited by Robert T. Gunther (Oxford; reprint New York, 1968).

DOSSIE, ROBERT. 1758. *The Handmaid to the Arts* (2 v., London).

DUCHESNE, L. 1886. *Le Liber Pontificalis I* (Bibliothèque des Ecoles Françaises d'Athenes et de Rome, 2e serie, Paris).

DÜWEL, KLAUS. 1968. *Runenkunde* (Stuttgart).

EASTLAKE, CHARLES L. 1847, 1869. *Materials for a history of oil painting* (2 v., London).

EDELSTEIN, S. M., and H. BOGHETTY. 1965. "Dyeing and Tanning Leather in the XVth century." *Amer. Dyestuff Reporter* **54**: pp. 940–944.

EDELSTEIN, S. M. 1963. "Dyestuffs and dyeing in the XVth century." *Amer. Dyestuff Reporter* **52**: pp. 2–5.

EICHHOLZ, D. E. 1965. *Theophrastus de Lapidibus. Edited with translation and commentary.* (Oxford).

ERACLIUS. *De coloribus et artibus Romanorum.* For text of this see Merrifield (1849).

ERCKER, LAZARUS. 1574. *Beschreibung allerfürnemisten mineralischen Ertzt und Berckwercksarten* (Prague). Fifth edition, with extensive notes by Johannes H. Cardalucius, *Aula Subterranea . . . das ist Untererdische Hofhaltung . . .* (Frankfurt, 1672). English translation by A. Sisco and C. S. Smith (Chicago, 1951).

FERGUSON, JAMES. 1888. "Some Early Treatises of Technological Chemistry." *Proceedings of the Philosophical Society of Glasgow* **19**, pp. 126–159; and supplements, 1894; **25**: pp. 224–235; 1911: **43**: pp. 232–258; and 1912: **44**: pp. 149–189.

FORBES, R. J. 1948. *Short History of the Art of Distillation* (Leiden).

GANZENMÜLLER, WILHELM. 1941. "Ein unbekanntes Bruchstück der Mappae Clavicula aus dem Anfang des 9 Jahrhunderts." *Mitt. zur Geschichte der Medizin der Naturwissenschaft und der Technik* **40**: pp. 1–15.

GETTENS, R. J., and G. L. STOUT. 1947. *Painting Materials: A Short Encyclopedia* (New York).

GETTENS, R. J., R. L. FELLER, and W. T. CHASE. 1972. "Vermilion and Cinnabar." *Studies in Conservation* **17**: pp. 45–69.

GIRY, A. 1878. "Notice sur un traité du moyen-âge intitulé *De coloribus et artibus Romanorum,*" *Bibliothèque de l'Ecole [pratique] des Hautes Etudes*. 35° fascicule; mélanges publiés par la section historique et philologique de l'Ecole des Hautes Etudes pour le dixième anniversaire de sa fondation (Paris).

GRABAR, ANDREI N. 1957. *Early Medieval Painting from the Fourth to the Eleventh Century. Mosaics and Mural Painting by A. Grabar, . . . Book Illustration by Carl Nordenfalk*, translated by S. Gilbert (Lausanne).

GÜNTHER, SEBASTIAN. 1810–1815. *Geschichte der Literaturischen Anstalten in Bayern* (Munich) **1**: p. 397.

HEDFORS, HJALMAR. 1932. *Compositiones ad tingenda musiva* [Codex Lucensis 490] *Herausgeben übersetzt und philologisch erklärt . . .* (Uppsala).

HERO OF ALEXANDRIA. 1899. *Opera*, Wilhelm Schmidt (Editor), Vol. 1, *Pneumatica et automata* (Leipzig,). The appendix to Vol. 1, pp. 458–489 contains, in Latin and German trans., fragments of Philo *Pneumatica* including the *Liber Philonis De Ingeniis Spiritualibus*.

ILG, ALBERT. 1874. *Quellenschriften für Kunst-geschichte und Kunsttechnik des Mittelalters und der Renaissance* (Vienna).

JOHANNSEN, OTTO. 1933. "Gab es in der Karolingerzeit schon Hochofen." *Stahl und Eisen* **53**: pp. 1039–1040.

JOHNSON, ROZELLE PARKER. 1935. "Note on Some Manuscripts of the *Mappae Clavicula*." *Speculum* 10: pp. 72–76.

——. 1935. "Additional Notes on Some Manuscripts of the *Mappae Clavicula*." *Speculum* 10: pp. 76–81.

——. 1937. "Some Continental Manuscripts of the *Mappae Clavicula*." *Speculum* 12: pp. 84–103.

——. 1939. *Compositiones variae . . . . An introductory study* (*Illinois Studies in Language and Literature* 23, 3, Urbana).

KÜHN, HERMANN. 1968. "Tin-lead Yellow." *Studies in Conservation* 13: pp. 7–23.

——. 1970. "Verdigris and Copper Resinate." *Studies in Conservation* 15: pp. 12–36.

LAGERCRANTZ, O. 1913. *Papyrus graecus Holmiensis. Recepte für Silber, Steine, und Purpur* . . . (Uppsala and Leipsig).

LE BEGUE, JEHAN. For text and translation of the manuscripts compiled by Le Begue in 1431 [BN6741], see Mary P. Merrifield 1849 1: pp. 1–319.

LECHTMAN, HEATHER. 1973. "The Gilding of Metals in pre-Columbian Peru," in W. J. Young, editor, *Application of Science in the Examination of Works of Art* Museum of Fine Arts, Boston pp. 38–52.

LEWIS, WILLIAM. 1763. *Commercium Philosophico-Tecknicum or Philosophical Commerce of Arts* (London).

LIPPMANN, E. O. VON. 1890. *Geschichte des Zuckers* (Leipsig).

——. 1920. "Zur Geschichte des Alkohols." *Chemiker-Zeitung.* **44**: p. 625.

LOWE, ELIAS AVERY. 1934–1966: *Codices Latini Antiquiores* (11 v., Oxford).

[MACQUER, PHILLIPE]. 1761. *Dictionnaire portative des arts et metiers* (2 v., Paris).

MAIGNE D'ARNIS, W. H. 1858. *Lexicon manuale ad scriptores mediae et infimae latinatis* . . . (Paris).

MERRIFIELD, MARY PHILADELPHIA. 1849. *Original treatises, dating from the XIIth to XVIIIth centuries on the arts of painting in oil* (2 v., London).

MIELI, A. 1939. *La science Arabe et son rôle dans l'évolution scientifique mondiale* (Leyden).

MUHLETHALER, B., and J. THISSEN. 1969. "Smalt." *Studies in Conservation* 14: pp. 47–61.

MUNBY, ALAN NOEL LATIMER. 1951–1960. *Phillipps studies* (5 v., Cambridge).

No. 1—*The catalogues of manuscripts and printed books of Sir Thomas Phillipps*, 1951;

No. 2—*The family affairs of Sir Thomas Phillipps*, 1952;

No. 3—*The formation of the Phillipps library up to the year 1840*, 1954;

No. 4—*The formation of the Phillipps library from 1841–72*, 1956;

No. 5—*The dispersal of the Phillipps library*, 1960.

—— (adapted by Nicolas Baker). 1967. *Portrait of an obsession. The life of Sir Thomas Phillipps, the world's greatest book collector* (London).

MURATORI, LUDOVICO ANTONIO. 1738–1742. *Antiquitates Italicae medii aevi* . . . (6 v., Mediolani). Dissertatio XXIV, Vol. II (1739), cols. 349–396. "De Artibus Italicorum post inclinationem romani imperii." The Latin text of the Lucca *Compositiones variae*, which appears as part of the *Liber Pontificalis*, occupies cols. 365–388.

NEEDHAM, JOSEPH. 1965. *Science and Civilisation in China* 4, 2, *Mechanical Engineering* (Cambridge).

NORDENFALK, CARL. See Grabar, A. N. (1957).

OPPENHEIM, A. LEO, *et al.* 1970. *Glass and Glassmaking in ancient Mesopotamia* (Corning, N. Y.)

PARTINGTON, JAMES R. 1953. *Origins and Development of Applied Chemistry* (London).

——. 1960. *A History of Greek Fire and Gunpowder* (Cambridge).

PHILO OF BYZANTIUM. *De Ingeniis Spiritualibus.* Text and translation given in Vol. 1 of the W. Schmidt edition of the *Opera* of Hero of Alexandria, *q.v.*

PHILLIPPS, SIR THOMAS. 1847. "Letter addressed to Albert Way, Esq., Director, communicating a transcript of a MS treatise on the preparation of pigments, and on various processes of the decorative arts practiced in the Middle Ages, written in the twelfth century, and entitled *Mappae Clavicula*." *Archaeologia* 32: pp. 183–244.

PLESTERS, JOYCE. 1966. "Ultramarine Blue, Natural and Artificial." *Studies in Conservation* 11: pp. 62–91.

PLINY (C. Plinius Secundus). *Historia naturalis.* The chemical sections are best read in K. C. Bailey's edition, 1929–1932, *q.v.*

PORTA, GIOVANNI BATTISTA DELLA. 1589. *Magiae naturalis libri viginti* . . . (Naples). Anonymous Eng. trans. (London, 1658; reprinted 1669, 1957).

RAFT, A. 1968. "About Theophilus' Blue Color, 'Lazur.'" *Studies in Conservation* 13: pp. 1–6.

RÉAUMUR, R. A. F. DE. 1741. "L'art de faire une nouvelle sorte de procelaine, . . . ou de transformer le verre en porcelaine." *Mem. Acad. Sci., 1749*, pp. 370–388.

ROOSEN-RUNGE, HEINZ. 1967. *Farbgebung und Technik frümittelalterlicher Buchmalerei. Studien zu den Traktaten "Mappae Clavicula" und "Heraclius"* (2 v., Munich, Deutscher Kunstverlag).

——. 1970. "Farben-und Malrezepte in frühmittelalterlichen technologischen Handschriften." In E. E. Ploss, H. Roosen-Runge, H. Schipperges and H. Buntz, *Alchima: Ideologie und Technologie* (Munich), pp. 47–66.

——. 1972. "Die Tinte des Theophilus." In *Festschrift Luitpold Dussler* (Munich, Deutscher Kunstverlag), pp. 87–112.

ROSETTI, GIOANVENTURA. 1548. *Plictho de larte de tentori* . . . (Venice). Facsimile reproduction and English translation by S. M. Edelstein and H. C. Borghetty (Cambridge, Mass., 1969).

SANTILLANA, GIORGIO DE, and HERTHA VON DECHEND. 1969. *Hamlet's Mill: An Essay on Myth and the Frame of Time* (Boston, Mass.).

SCHIAPARELLI, LUIGI. 1924. *Il codice 490 della Biblioteca capitolare di Lucca Ottantatre pagine per servire a studi paleografici . . . reprodotte in fototipia* . . . (Rome).

SCHMIDT, WILHELM (Editor). See Hero of Alexandria.

——. 1924. *Il codice 490 della Biblioteca Capitolare di Lucca Ottantatre pagine per servire a studi paleografici . . . reprodotte in fototipia* . . . (Rome).

Secrets concernant les arts et metiers (4 v., Paris, 1724).

SMITH, CYRIL S. 1969. "Porcelain and Plutonism." In *Toward a History of Geology*, Cecil J. Schneer, editor (Cambridge, Mass.).

——. 1960. *A History of Metallography* (Chicago).

——. 1970. "Art, Technology and Science: Notes on their Historical Interaction." *Technology and Culture* 11, pp. 493–549. Reprinted with commentary in D. H. D. Roller, editor, *Perspectives in the History of Science and Technology*, (Norman, Oklahoma, 1971) pp. 129–176.

——. 1972. "Metallurgical Footnotes to the History of Art." *Proc. Amer. Philos. Soc.* 116, 2: pp. 97–135.

——. 1974. "Historical Notes on the Coloring of Metals" in Adli Bishay, editor, *Proceedings of the Second International Conference on Solid State Physics, Cairo, 1973* (New York).

SMITH, GODFREY. 1740. *Laboratory or School of Arts* (London).

SVENNUNG, J. 1941. *Compositiones Lucenses. Studien zum Inhalt, zur Textkritik und Sprache* (Uppsala Universitats Ärsskrift, No. 5., Uppsala and Leipsig).

THEOBALD, WILHELM. 1933. *Technik des Kunsthandwerks im zehnten Jahrhundert des Theophilus Presbyter Diversarum artium schedula, in auswahl neu herausgegeben, übersetzt und erlaütert* . . . (Berlin).

——. 1912. "Die Herstellung des Blattmetalls in Altertum und Neuzeit," *Glasers Annalen für Gewerbe und Bauwesen* (Berlin) **70**: pp. 91–99.

THEOPHILUS. *De diversis artibus.* Manuscript *ca.* 1123 A.D Latin text and English translation by C. R. Dodwell (London, 1961). English translation with technical notes by J. G. Hawthorne and C. S. Smith (Chicago, 1963).

THEOPHRASTUS, *Peri Lithon.* See Caley, E. R. and J. C. Richards (1956), and, best edition, D. E. Eicholz (1965).

THOMPSON, DANIEL V. 1936. *The Materials of Medieval Painting* (London).

THOMPSON, DANIEL, V. 1967. "Theophilus Presbyter: Words and Meaning in Technical Translation." *Speculum* 62: pp. 313–339.

THOMPSON, R. C. 1936. *A Dictionary of Assyrian Chemistry and Geology* (Oxford).

THORNDIKE, LYNN. 1923–1958. *A History of Magic and Experimental Science* (8 v., New York). The *Mappae Clavicula* is discussed in Vol. 1, chap. 33.

THUROT, CHARLES. 1868–1869. "Recherches historiques sur le principe d'Archimede." *Revue Archaeologique* 18: pp. 389–406; 19: pp. 42–49, 111–123, 284–299, 345–360; 20: pp. 14–33.

URE, ANDREW. 1842. *Dictionary of Arts and Manufactures* (London and New York). (This is an enlarged version of a work that was first published in 1821, as a revision of the still earlier *Dictionary of Chemistry* by Wm. Nicholson. It gives an excellent internal view of the traditional arts at a time when chemical science could explain them but had not yet changed their nature and scale.)

WHITE, LYNN. 1962. *Medieval Technology and Social Change* (Oxford).

INDEX

Note: This index is partly interpretative and occasionally refers to topics in modern terminology, using words that will not be found at the place cited.

The numbers in roman type are references to chapter numbers, with n denoting a footnote. Italic type, which is used only for items in the introduction and appendices, refers to pages.